Charles A. Bunting

Hope for the Victims of Alcohol, Opium, Morphine, Cocaine, and Other Vices

Charles A. Bunting

Hope for the Victims of Alcohol, Opium, Morphine, Cocaine, and Other Vices

ISBN/EAN: 9783744670661

Printed in Europe, USA, Canada, Australia, Japan

Cover: Foto ©berggeist007 / pixelio.de

More available books at **www.hansebooks.com**

HOPE

FOR THE VICTIMS OF ALCOHOL,

OPIUM, MORPHINE, COCAINE, AND OTHER VICES.

A NARRATION OF

SUCCESSFUL EFFORTS

DURING

TEN YEARS OF PERSONAL LABOR,

DEVOTED AS CHRIST'S INSTRUMENT

TO REDEEM THE SLAVES OF SUCH HABITS,

IN THE

◄ NEW YORK CHRISTIAN HOME ►

FOR

INTEMPERATE MEN.

(With Opinions on Moderate Drinking, High License, the Rum Traffic, etc.)

BY

CHARLES A. BUNTING.

With full page Photo-engravings of the late W. E. Dodge: of J. L. Pu...
Author: of the old Home and of the New Ho...

NEW YORK
CHRISTIAN HOME BUILDI...
1175 MADISON AVE.,
1888

COPYRIGHT,
1888.
BY CHARLES A. BUNTING.

TO HER

IN WHOM I REALIZE THAT

"WHOSO FINDETH A WIFE FINDETH A GOOD THING,
AND OBTAINETH FAVOUR OF THE LORD;"

TO THE HELPMATE

WHO HAS SORROWED IN MY GRIEFS AND REJOICED IN MY JOYS.

TO MY WIFE

WHO HAS BORNE MANY A CROSS THAT I MIGHT WEAR A CROWN,

AND OF WHOM I CAN BEAR TESTIMONY THAT

"HER PRICE IS FAR ABOVE RUBIES.".

THE FOLLOWING PAGES ARE DEDICATED.

WM. E. DODGE.

PREFACE.

The object of this little work is self-explanatory, and I have no reason to apologize for presenting it to those who are necessarily interested on either side of the great standard upon whose pure folds is emblazoned the simple but significant legend—" Total Abstinence." I cannot offer apology to those who are either actively engaged in the service of the Demon of Drunkenness, or else look supinely on while they every day see their brothers brought to the pit of mental, moral, physical and spiritual destruction. To either of these classes my feeble words but witness against their sins of commission and omission, without moving their souls to abandon the banner of Satan and put on the armor of Christ. Nor have I apology to offer those who are, and have been, engaged in the glorious crusade of Gospel Temperance. They realize the evils which beset us on every hand and recognize the value of every blow struck at the degrading, debasing habits which becloud the nobler portion of man's physical existence, and change the bodies, made to be temples of the Holy Ghost, into cesspools of filth and iniquity. Hence with all of its shortcomings, I send this volume forth, bespeaking for it solely the calm consideration of thoughtful, earnest men and women.

Without undue presumption on the one hand, while avoiding mistaken modesty on the other, I can fairly claim that in dealing with this question I know whereof I speak. An experience of over ten years' duration justifies me in believing that in all which is connected with the great work of reforming the intemperate or of combating the agencies employed by the Arch-enemy of Christ to lure the unsuspecting and unfortunate into the byways of sin, I have no uncertain and, assuredly, no unascertained place. Palter with drunkenness as you may, tinker with the license system as you will, yet certain as that God lives, that heaven and earth shall pass away, but that His word shall abide forever, so it is that apart from the following of Christ Jesus, there is neither prevention nor cure for the sin of intemperance.

I commend the facts detailed in these pages to the Christian philanthropist, the Christian teacher, the Christian lawmaker. In all kindness they are also commended to the attention of those about to start on the path of life, hoping that this cry of warning may not fall on wholly unheeding ears, and especially are they narrated for the sake of those who have fallen, who have gone astray, who are among the lost sheep whom it is the delight of Jesus to seek and to save.

I would here, before finally committing these pages to my considerate readers, acknowledge the practical assistance rendered me in their preparation by my dear young friend, Mr. Thomas H. Kilduff, and that he may ever feel the blessed peace of the righteous disciple is my grateful prayer.

And now, rendering all glory to God, I invoke a blessing on all who labor in this holy work, confident that as His providence reacheth from end to end, disposing of all things sweetly, so will he give plentiful increase to the seed which has been planted in His name and sprinkled with the blood of His beloved Son.

CHARLES A. BUNTING.

New York, June 7, 1887.

TEN YEARS' WORK IN THE
NEW YORK
CHRISTIAN HOME.

ORIGIN OF THE WORK.

At the earnest request of my many friends and those interested in the great work of saving the drunkard, I herein give a brief history of the origin and work of The New York Christian Home for Intemperate Men.

To God I feel grateful for the grace which has enabled me to meet and overcome the obstacles that presented themselves in this untried and, to many, doubtful undertaking.

It is pleasurable to make record of how Christ has again shown that our extremity is His opportunity, and to present to you the facts and figures of that which has been, through His aid, a decade of years of blessing and success.

In the month of February, 1876, my attention was directed in a most peculiar manner to my own life. Reared as I was under the most advantageous circumstances for entering a Christian life in my youth, the great and most important of all things, the salvation of my soul, was neglected. But during a series of meetings held in this city in the winter and spring of 1876, under the evangelist, Mr. D. L. Moody, I sought and found rest by believing on Jesus as my Saviour, and in the Fall of that year I became a member of the Church of the Holy Trinity, Rev. Stephen H. Tyng, Jr., D.D., rector.

During the Fall and Winter of 1876 and 1877 a series of revival meetings was held in our church and I became very active in them. As an outgrowth of these meetings a noon-day prayer-meeting was held daily in the basement of the church; also a Gospel temperance service every Sunday afternoon in the chapel of the church, both of which were under my special management.

At the close of these Sunday afternoon meetings a supper was provided under the supervision of the assistant, Rev. Wm. Humpstone, for the unfortunates who had gathered in the chapel. The plan proved so successful as to the number present at this temperance service that before the close of the winter we had an average attendance at these suppers of over five hundred men. At these meetings and suppers I was constantly brought in contact with great numbers of helpless and apparently hopeless inebriates, men who had fallen so low that they had lost situation and friends, reputation and character; and yet in many instances I became convinced that if there was some place of refuge where they could be temporarily cared for and kept from the corrupt surroundings of the cheap lodging houses, station houses, etc., some at least of the many who professed a willingness to begin a new life could be saved. This thought led to many interviews with my pastor, Rev. Dr. Stephen H. Tyng, Jr., as to the advisability of opening a temporary home and refuge for the restoration of these men. Many plans were suggested, among which was one to lease a loft, fit it up with temporary bunks, and provide not only lodgings but a breakfast, say of coffee and rolls, the place to be in charge of some person capable of handling such a class of men. While this plan was being agitated I was not idle, but wrote many letters to the benevolent people of this city seeking encouragement and aid for the support of such a place and the maintenance of a suitable person to conduct it.

Early in February, 1877, nothing definite having been decided on up to that time, my mind became more impressed than ever with the importance of establishing this work, but as my former efforts had been of no avail I was at a loss to know how best to proceed.

Believing as I did that all reliance for success in the recovery of lost men should be based upon the application of the gospel of Christ, I again appealed to God for guidance and again made the whole matter a subject of special prayer.

It was then for the first time I became impressed that I myself might be the man that God purposed to use in this work. On the sixth day of May, 1877, I asked God in prayer to direct me at once as to my duty in the work which I so forcibly felt was laid upon me, and I shall never forget how soon my prayer was answered, for in the words of Isaiah I could exclaim—" And it shall come to pass, that before they call I will answer. And while they are yet speaking I will hear." As I arose from my knees the door-bell rang. It was our late worthy President, the Hon. William E. Dodge, who had called to see me to ascertain whether there was any place in the city where a man, sick and tired of a life of sin and broken down by his intemperate habits, might find a resting place and be shut off for a time from his old customs and associations. I quickly communicated to Mr. Dodge the thoughts that were then upon my mind, and told him how desirous I was to be led by the spirit of God in this great undertaking and how I had been asking to be directed in the matter by Him. As soon as I had finished this conversation Mr. Dodge assured me that it was truly the spirit of Christ that had prompted me and that I should have his prayers and help in the work ; and although I was at this time under several business contracts, the relinquishing of which would result in considerable pecuniary loss to me, I decided at once to give up everything and to devote my life to this work in which the Lord had undoubtedly called me.

On the following day, May 7th, I called on Mr. R. R. McBurney, the Secretary of the Young Men's Christian Association, to submit my plan to him. He thought at first, among other objections, that it was too late in the season to think of starting such a work, as many among those to whom we would have to look for support would soon leave the city for country homes ; but when I showed him the prospectus that I had drawn up and which Messrs. Wm. E. Dodge and John B. Gough had endorsed and agreed to support, he said "Go on." The following is the prospectus I submitted to Mr. McBurney :

" It is proposed to open a temporary home and refuge for the restoration of men who have fallen victims to strong drink. A suitable house will be selected and conducted under the charge of C. A. Bunting. A limited number of those needing Christian help will be admitted. The administration of the Home will be domestic. All reliance for success in the recovery of lost men will be based upon the application of the Gospel of Christ and the restraints of Christian example and fellowship.

The Home will be administered by a Board of Managers representing the different denominations of the church.

Its projector is authorized to say that his efforts have our cordial commendation and support.

"Signed:
"WILLIAM E. DODGE,
"JOHN B. GOUGH."

The next day I received the following letter:

Young Men's Christian Association of the City of New York,
Twenty-third Street and Fourth Avenue.

My Dear Mr. Bunting,—I have had a talk with one of the best men in town about your matter this evening. He is deeply interested. Come and see me before 10 to-morrow if convenient. I am yours,

R. R. McBURNEY.

8th May, '77.

OUR FIRST MEETING.

On my calling upon Mr. McBurney he gave me a letter of introduction to Mr. William T. Booth, the gentleman he referred to in his communication, and that evening or the evening following I had an interview with Mr. Booth in which it was agreed to call a meeting at the rooms of the Young Men's Christian Association of persons likely to be interested in the matter. The first meeting held to take into consideration the carrying out of my plans was convened in the parlors of the Y. M. C. A. on Friday evening May 18, 1877.

There were present in that meeting the following gentlemen: Messrs. Caleb B. Knevals, Rev. D. Stuart Dodge, Wm. T. Booth, R. R. McBurney, Arthur W. Parsons, and myself.

On motion of Mr. McBurney, Mr. W. T. Booth was called to the chair, and Mr. Arthur W. Parsons was appointed Secretary. After prayer by Rev. Mr. Dodge, the Chairman stated the cause for calling the meeting. I was then invited to give in full my ideas on the need of a home for the intemperate, with my plans for conducting such a work, and when I had fully unfolded my scheme it received the hearty approval of those present. At once Mr. Knevals urged the advisability of the immediate establishment of such an institution, the want of which had never occurred to him before. Rev. D. Stuart Dodge offered the following preamble and resolutions, which were seconded by Mr. McBurney and passed unanimously :

Whereas, Large numbers of men desiring to be delivered from intemperate habits are continually met in religious meetings or are brought to the attention of those engaged in religious work ; and,

Whereas, It is our conviction that instead of treating this class chiefly as diseased and needing medical care or requiring forced restraint they should be led to look to God above for aid and to feel that no true or permanent change of life can be hoped for except in a change of heart, in strength derived from distinctly Christian principles ; and,

Whereas, There exists no Christian institution where such persons can temporarily receive the peculiar sympathy and experienced aid their circumstances demand and be brought under direct religious influences, therefore

Resolved, That steps be taken to secure for this purpose a house suitably located at a moderate rent and to be conducted with the strictest economy, and that for the organization and control of the same the following gentlemen be named to constitute the Board of Managers : William T. Booth, J. Noble Stearns, Samuel C. Burdick, John S. Bussing, Edmund Penfold, Arthur W. Parsons, Caleb B. Knevals, J. M. Cornell, James Talcott, Charles A. Bunting, and D. Stuart Dodge.

These gentlemen were elected.

Another meeting was held May 22, 1877. Reports were made, and after discussing many plans for raising means to defray the necessary expenses until a public meeting could be held after the summer months, it was

Resolved. That Mr. Dodge, Mr. McBurney, and Mr. Parsons be appointed a committee with power to draw up an appeal to the public, and that this appeal be sent for signatures to the gentlemen proposed as Board of Managers. That this appeal be published in the daily and religious papers of this City after being signed by the Chairman and Secretary of the meeting, of which the following is a copy :

A PRACTICAL WORK.

Efforts to restore the intemperate show that numbers of hopeful cases are continually lost from want of a suitable place in this city where men, sincerely anxious to reform, can find the encouragements and restraints needed in their first struggles to resist a depraved appetite and overcome the effects of entire cessation from the use of stimulants.

It is now proposed to provide a house where such persons can find a temporary refuge, and where especially they may be brought under direct religious influences.

All able to pay will be expected to meet their own expenses, but none who give evidence of an honest desire to be reclaimed will, if possible, be excluded. In the management of the house regard will be had for the strictest economy and oversight. An earnest appeal is made for the funds to enable the committee to assume the responsibility of renting a building and meeting necessary expenses.

Contributions or applications for admission may be sent to any one of the undersigned.

WILLIAM T. BOOTH, Chairman.		ARTHUR W. PARSONS, Secretary,
SAMUEL C. BURDICK,	EDMUND PENFOLD.	JAMES TALCOTT.
JOHN N. STEARNS.	CALEB B. KNEVALS,	J. MILTON CORNELL,
D. STUART DODGE,	WILLARD PARKER, M.D.,	CHARLES A. BUNTING.

SECURING A LOCATION.

On the 31st of May, 1877, another meeting was held, at which were present Messrs. Burdick, Talcott, Knevals, Penfold, D. Stuart Dodge, Parsons, Bunting, Bussing, and Stearns. Mr. Knevals being called to the chair, requested Mr. Dodge to open the meeting with prayer. Mr. Burdick moved that hereafter all the meetings of this committee be opened with prayer, which was duly seconded and carried. On motion of Mr. Burdick it was resolved that the Chairman appoint two sub-committees of three each, the first to report a plan for permanent organization *if such were deemed desirable*, the second to look into the cost and location of a proper building, and also the cost of continuing the work six months from June 1st. The Chairman appointed as the first of the sub-committees, Messrs. Booth, Burdick, and Penfold, as the second Messrs. Dodge, Parsons, and Bunting.

In conversation with Dr. Tyng about this time he kindly offered to give me a letter to a friend of his who was desirous of disposing of the lease of a house, and upon making an examination it proved to be exactly what I wanted and adapted in every particular to the needs of the work. I at once secured the refusal of it, as well as of the carpets, gas fixtures, etc., which were at that time in it.

At a subsequent meeting held June 4th, Mr. Dodge for his committee, reported that the probable expenses for running the Home for one year with thirty members, including furnishing of the same would be about $7,000. Mr. Burdick for committee on organization, reported as follows: "Your committee recommend that for the present the efforts to establish a Home, etc., be conducted under the charge of a general committee consisting of twelve members, including Chairman, Secretary and Treasurer," which was adopted.

On motion of Mr. Knevals, William T. Booth was elected permanent Chairman, James Talcott, Treasurer, and Arthur W. Parsons, Secretary. At this meeting a house committee was appointed and authorized to hire the house No. 48 East 78th Street, of which I had previously secured the refusal.

THE HOME OPENED.

This house capable of accommodating about thirty men and well adapted for the purpose, was soon put in readiness for occupancy. Through the kindness of one of the committee, a son of our late President, Mr. Dodge, who was deeply interested in the movement, several of the rooms were furnished, the balance necessary to completely apparel the house being purchased, and on the 7th day of June 1877, The New York Christian Home for Intemperate Men, was opened.

At an informal meeting held in August for the purpose of "Organizing under the General law," Mr. Parsons reported that it was possible to organize and at once. It was thereupon voted that this same Committee be continued and that they take steps for immediate incorporation under the General law. At a meeting held October 17th, 1877, Mr. Parsons reported, and submitted the necessary certificate of those who were ready to subscribe their names as incorporators. (copy of the act of Incorporation, and the subsequent act of Reincorporation—See supplementary matter.) At this meeting it was agreed upon that the number of Directors for the first year be thirteen, and the following gentlemen were named to constitute the Board of Directors: William T. Booth, Arthur W. Parsons, James Talcott, Caleb B. Knevals, Samuel C. Burdick, Robert R. McBurney, Edmund Penfold, William E. Dodge, John Noble Stearns, Rev. D. Stuart Dodge, Chas. A. Bunting, Willard Parker, M. D., Washington R. Vermilye.

The first meeting of the Board of Directors was held Oct. 19, 1877.

Mr. Stearns moved that the permanent officers of the Society for the year be a President, Vice-President, Secretary, Treasurer and Resident Director. Mr. William T. Booth was elected President, Mr. Wm. E. Dodge, Vice-President, Mr. Arthur W. Parsons, Secretary, Mr. James Talcott, Treasurer and Mr. Charles A. Bunting, Resident Director.

A committee was appointed at this meeting to draw up the Constitution and By-laws by which the organization should be governed.

Nov. 1st, 1877 the By-laws and Constitution were presented and after reading they were adopted. (See supplementary matter.)

OBJECT AND PLAN OF WORK.

I will now state as briefly as possible the exact object, plan of work, etc., which were adopted at the outstart.

The object of the Home was not to seek the cure of the intemperate by medical treatment nor was it intended to be an asylum for permanent residence. It was opened with the full assurance that though such men might be helped by mere moral or physical agencies, such agencies were insufficient and could not of themselves save them. We seek to impress upon the drunkard who comes under our care, the fact that drunkenness is a sin against God, to be repented of and forsaken, therefore, the Gospel Remedy is applied to the alcoholic and narcotic habits, which are regarded and treated purely and simply as sins.

Any applicant however wretched or low his condition, is entitled to a kindly greeting and a patient hearing. If it is reasonably evident that he comes with an honest purpose to abandon his evil habits, he is admitted. Men in this plight need a retreat where they can agonize mentally and physically in "the valley of decision." They need repose in order to gain strength. They need kind and careful treatment to bear them up under the prostration which the demon alcohol or the fiends of opium, morphine or cocaine have produced. Above all they need to be cut off from the scenes, persons and associations of the unhappy past. The Home offers these aids to reformation. Whoever enters it leaves alcohol, opium and morphine behind, for there is no tapering off, as it is called; the severance is absolute, immediate and complete.

Our course in this respect has not been without criticism, and severe criticism at that; but the fact must never be lost sight of that it is upon God we rely, and that His hand is in the work can be no better illustrated perhaps than by the fact that out of the hundreds of men that we have thus from time to time cared for, but three deaths have occurred in our midst. He has certainly led us on. Amidst all trials and discouragements we have heard His voice saying "Fear not, for I am with thee: be not dismayed, for I am thy God; I will strengthen thee: yea I will help thee, yea I will uphold thee by the right hand of my righteousness."

And the fact above referred to is perhaps made more surprising, when I tell you that among those received have been men who had reached a daily allowance of, in many instances, from 100 to 200 grains of opium and up to 60 grains of morphine.

To restore the sufferer to physical health, the management confidently rely upon the recuperative powers of nature, aided by nutritious food, regular hours, open air exercise, with happy surroundings, and this confidence is based upon an observation of almost invariable successful treatment.

In case a patient is under the influence of alcohol when he comes to us, especially if he need to be placed in our temporary hospital, he is treated with some ordinary sedative. Other medical attendance is rarely required.

After two or three days, unless they be opium or morphine patients, they are allowed the freedom of the Home.

Inmates who have means of their own, or friends able to help them, are expected to pay according to the accommodations provided; but all alike, rich or poor, come under the same regulation respecting recourse to drugs or palliatives. Abundant and nutritious food is supplied; the quiet and comforts of a true home are enjoyed, and the temptation or opportunity to drink is absolutely removed: but our constant and chief reliance is upon moral motives, vitalized by Christian principles.

HELPING ONE ANOTHER.

The Institution was established for those only who seriously desire to be saved and who are ready to look to God for power to overcome their degrading appetite. Among the most positive and fruitful agencies in the Home is the influence of the members upon each other. As one after another gets his feet upon the Rock, he naturally and gladly turns round to stretch out a hand to those struggling to gain a place beside him. A spirit of tender sympathy and helpfulness thus pervades the entire circle, not only strengthening the resolutions of every new convert, but preparing them all for similar efforts when they have left the Home. It is gratifying to know that a large number of those who go out from this companionship, engage at once in some form of Christian work, not as a means of support, but as an expression of gratitude and from a desire to lift up others out of the wretchedness they themselves once knew. Men who have been for twenty, thirty, and in some instances for over forty years, habitual spirit drinkers and inveterate smokers, are enabled after a few weeks' residence in the Home to "lay aside every weight and the sin which doth so easily beset," and relying upon God's grace, return to their usual avocations; the clergyman to his pulpit, the lawyer and physician to their practice, and the merchant and salesman to the "receipt of customs." They become evangelists of temperance, and carry, each to his own circle of associates, the saving power of the Gospel and the good news that recovery is indeed possible. Freedom from restraint is one of the distinctive features of the Home, although it is one of the rules, that all inmates are expected to remain in the house, until the manager or his assistant, is satisfied that they can be allowed in safety to take their regular walk with the other members: and all arrangements for going out regularly, or occasionally, must be made with the Resident Manager. This freedom from restraint, and the sense of personal liberty, combined with the appeals which are constantly made to a man's own conscience, honor, self respect, and higher nature, are found to be better than the coercive system of bit and bridle. To awaken this higher nature, and to bring a man to himself, is the one supreme object of the Home.

DRUNKENNESS NOT HEREDITARY.

To palliate the drunkard's sin by acknowledging a belief in such a sin being hereditary (even if that be a palliation) is no kindness; it is only encouraging him in self-deception, which is the great obstacle to repentance and recovery.

We do not believe drunkenness is inherited any more than any other sin or evil; therefore the inmate has his mind disabused of that idea by putting the responsibility where it rests, alone upon himself. Of 293 members received in the Home for one year, 52 only were the offspring of intemperate parents while 197 claimed association as the cause of their drinking, and this law of association that had controlled in their ruin, is the same law that we invoke under God for their rescue.

The cases of moral victory which have occurred in the Home seem almost incredible to those who have not been witnesses. An earnest and talented gentleman whom the opium habit had demented and rendered incapable of taking care of his family or of attending to the active duties of life, had reached the daily allowance of two hundred grains. He was admitted to the Home, sought refuge in the Everlasting Arms, and in a moment abandoned the opium wholly, and he has never touched it since.

His last testimony here was, that he was entirely free from the habit, never having had a desire to touch it since he gave his heart to God. If the opium habit can be thus vanquished in a day by the regenerating power, certainly the habit of drinking can be cured in those who have earnest determination and seek help from God.

No one is allowed to remain who does not, within a proper time manifest willingness to listen to such counsel and instruction. We have no desire to force our religious belief upon anyone, nor do we ask any to adopt a special creed or denominational system; but we do expect all to be present at the religious exercises, and to be honest in their efforts to break away absolutely and permanently from their evil habits. This Home was never designed for men who want an agreeable refuge in which to recover from one debauch and prepare for another. Immediately upon coming to the Home, a form of questions is asked the applicant, *one of which is particularly emphasized*—" Do you earnestly desire to permanently reform and become a Christian man ?"

Our work being personal, each man is taken into a room by himself, and there by earnest teaching and prayer we try to lead him to Christ. On Tuesdays we hold regular evening meetings expressly for the inmates of the Home, where those who profess to have found Christ are expected to testify.

After a suitable time has elapsed if we find there is no spirit of testimony or change in these men, and we can have no more religious influence over them, they are quietly informed that their places will be filled by others who are desirous of becoming followers of Jesus.

Our doors have always been open to all that are desirous of reformation, and I do not call to mind one single instance, since the opening of the Home, when we have turned away one that came to us with an honest purpose.

CONDITIONS OF ADMISSION.

The conditions on which we receive inmates are these. If a free bed is desired, the applicant or his friends must give satisfactory proof of his inability to remunerate the Home for his support during his stay therein ; all others will be expected to pay for their board weekly and according to their ability for the room, attendance, and accommodations furnished them. Another little safeguard we have thought advisable to use is that when a person seeks admission a paper is handed to him to be filled up by some one that he may know, as to his being a suitable person for admission. It reads as follows :

" We do hereby certify that we are personally acquainted with ―――― and believe him to be a fit party to be admitted to the benefits of The New York Christian Home for Intemperate Men." This, in many cases we have found to be very helpful to us, saving us frequently from imposition by those who were just following for the loaves and fishes. We are in constant communication with those that are absent from the city, so that we are able to give a report of about all that have ever been inmates of the Home.

NO NOSTRUMS ADMINISTERED.

In concluding this section of my narration I will add that the Board has not espoused the cause of any mere human theory of cure, either scientific, social, or moral. As Manager I have been, therefore, freed from all interest in nostrums and personal hobbies. Believing, as we do, that the great evil against which we contend is one of the almost innumerable forms of depravity, we seek its seat in the mind and soul rather than in the body, too often more sinned against than sinning. For such moral depravity the Gospel is the Divine specific.

The appliances employed in the application of this spiritual trusting and help are of the simplest, looking first to isolation for a little time from temptation, and when a condition of thorough sobriety has been secured, and the patient begins to find himself once more clothed and in his right mind (the matter of clothing frequently pertaining to the outer as well as the inward man), the motives to thorough and immediate reformation are urged upon him.

And here it is that the victory in most cases, perhaps we should say in all, has been won. The motives urged are distinctively Christian. The renewing of the heart is sought by suitable religious instruction, and earnest prayer to God for help. Social influences, the prospect of regained footing in the busy world, comfortable surroundings, and a wholesome diet—these are matters not overlooked; but it is due to truth to say that the supreme motives have been those which are urged upon all sinful men to repent and come to the foot of the Cross and be forgiven. The incentives of religion have ever been my strong reliance. And I might add, that no one responsibly connected with the Home would have its religious inculcations and influences any less prominent than they have been from the start; and this without implying any special theoretic accord on the mooted question as to the effects of conversion from vicious appetites.

As is apparent, it is necessary that each man be taken apart and then shown the sinfulness of his own heart, his necessity of appealing to God for aid and of complete reliance upon the promises found in His Word. This done it is easy to impress upon his mind that Christ is not merely a Saviour of mankind generally, but the Saviour of each distinct individual who receives him, remembering the words of our blessed Master, "Him that cometh to me I will in no wise cast out;" believing, also, "that He is able to save to the uttermost those who come unto God by Him."

In almost every instance the application for admission has been the voluntary choice of the sufferers. This fact secures for us as inmates men who are receptors of instruction and ready to yield to Christian influence.

It is but fitting that at this stage of the history of the Home, I speak of the self sacrificing and efficient gentlemen who have acted with me, some of them from the very commencement of the work. Mr. Booth our first President, filled that position from Oct. 19th, 1877 to Nov. 1st, 1880, (3 years) efficiently, and acceptably, when at his personal request his resignation was accepted.

I know I but voice the feelings of every heart when I say that in Mr. Booth we found a most zealous supporter of the Home, always ready by acts of devotion, and liberal donations, to bid us God speed. I can never forget the many kind words of sympathy and encouragement I received from him during the years of his Presidency: also the many kind, cheering words spoken to Mrs. Bunting throughout my sickness in the early part of the year 1878—called to fill as she was the trying position as the head of the large family. His resignation as President did not call for his withdrawal from the Board; and his name has been continued in our list of Directors from the outset, although much of the time, we have been deprived of his presence by sickness or absence from the city. Upon his resignation the Hon. Wm. E. Dodge, was elected to fill the vacancy.

Mr. John Noble Stearns was selected as Vice-President on the election of Mr. Dodge to the Presidency. Through him we came into possession of the valuable lots on which the Home now stands; he having purchased them for himself, but selling them to the Home at purchase price, when he could have disposed of them otherwise at a handsome profit. He acted also as a member of the BUILDING COMMITTEE, and his practical knowledge and experience, were of incalculable benefit to us in the erection of the building. From the time the Home was organized he was always a warm and hearty supporter of the work, giving liberally to the Building Fund, and contributing constantly, to meet current expenses.

Mr. Arthur W. Parsons who had filled the position of Secretary up to this date, on account of protracted illness was obliged to resign, tendering his resignation Nov. 16th. 1880. It was with feelings of deep regret that we were called upon to part with this able officer, so heartily interested was he in all that pertained to the Home work. His loss we felt not only as a devout Christian, but as a clear minded and practical adviser, ever ready with his talents and means to further the work, and it is with great pleasure I give an extract from one of his reports:

Secretary's Report for 1878-1879.

"In considering the work of The Christian Home for Intemperate Men during this second year of its existence, we have again to acknowledge the guiding hand of Providence. When our last report was issued, much had transpired to establish the confidence of those who were intimately acquainted with the work then so recently commenced.

"We now feel the Institution has proven its right to the support of all good citizens, and especially so of those who rely upon the power of the Divine Spirit to help them live according to the rules of Christian life and the dictates of an awakened conscience.

"The incidents and revelations which are continually coming to the knowledge of Mr. Bunting, who is the active worker with those for whom this Home was established, show that the power is at hand to help and equal to restrain even the most degraded, when such are ready to put themselves within reach of the Gospel message.

"It is when mingling their prayers with Mr. Bunting's that their hearts are so touched and softened that they cry for help, and acknowledge the only source from which they can receive assistance to restrain their evil appetite. As the bearer of Christ's message, Mr. Bunting's life is made glad in seeing so many brought to the conviction of the sinfulness of their lives, and on their entering the lines of Christian fellowship.

"The magnitude of the work done in the Home can not easily be estimated. Our numbers have been large. Still experience has compelled our house committee to limit the number of inmates—this limitation being dictated by considerations of health. A much larger number would readily come to the Home did space and accommodations permit. It is with the hope of meeting this demand for room, that some of our friends expect ere long to realize their anticipations of a house built for the permanent use of the Home. A committee has been appointed to have charge of the funds for such a building, and a considerable sum has been already pledged.

"A reference to the figures given in the statistical statement will show the numbers cared for and the cost of maintenance, and will prove that were such Homes established in all our cities they would be worthy of support for the sake of municipal economy, for it must be remembered that sooner or later most of those who follow the drunkard's course end as charges on the public purse. The economy can be understood when, as our tables show, sixty-five per centum of all who come to the Home become producers rather than consumers. Surely then may the question be asked, How much more worthy is such support when they become the homes of new men?—men who go forth clothed with the power of the Spirit and protected with the armor of righteousness?

"The public religious meetings have been held without intermission, and attended by large numbers, both from within and without the Home. To these meetings an invitation is extended to every one, and few who will attend them will fail to be convinced of the presence of God's Spirit.

"We are thankful that means have been provided so that the perplexity which last year attended all the actions of the society have been lessened. The increase in experience by all in charge has resulted in greater economy in the administration of the Home's family. We feel assured that all contributors will be satisfied of the economy exercised within the house, if they will consider carefully the number provided for and the amount of money expended.

"However, we would ask that this should not lessen the gifts of any. The manager's effort is to keep out of debt, but this can only be accomplished by contributions being received promptly and in sufficient amount."

Mr. Parsons died May 22, 1884, and I feel it a pleasure, as well as a duty, to say that as a man he was the most exemplary Christian it has ever been my lot to associate with in my Christian experience.

Mr. Matthew C. D. Borden succeeded Mr. Parsons, and proved an earnest supporter of the Home in many ways. Since his resignation the office has been filled efficiently by the present Secretary, Henry C. Houghton, M.D.

NEW HOME.

OUR FIRST HOME IN 78TH ST.

THE NEW HOME.

At the end of the third year the wants of our increasing family demonstrated that our quarters were insufficient to accommodate the numbers daily seeking admission—for, as many as we had cared for, we were forced from want of room to refuse an equal number. It had made our hearts ache to be compelled to say to a mother, pleading for the admission of her son, "We have no room." The increase of the work constrained us at this time to lay the facts of the case before Christian people, asking for means with which to build a house whose capacity would meet the wants now made apparent. The response to our call for funds was most noble, our subscriptions soon warranting the purchase of a site for our new Home. God opened the hearts of Christian men to give generously, and about the 1st of May, 1882, we were privileged to move into the new Home, which was formally opened and dedicated to God on the 11th of the same month. The building as represented in the accompanying illustration, occupies one of the finest and healthiest sites in New York, with the Central Park only one block distant, and is situated on the corner of Madison Avenue and 86th Street, with a frontage on the avenue of 100 feet and on the street of 38 feet, increasing in a portion of the structure to the depth of 52 feet.

In the cellar will be found a large drying-room, store-room, workshop, barber-shop, bath and toilet-rooms, coal-bins, and engine-room. On the first, or basement floor, the entrance to which is on 86th Street, are a reading-room, sitting-room, attaches' rooms, two dining-rooms, a laundry, kitchen and store-room. The first story is entered from Madison Avenue. Here is the office, a reception-room, a library and reading-room, a waiting-room, and a chapel, 24 ft. 6 in. x 50 ft. On the same floor are a dining-room and a large sitting-room. The private entrance for the family of the Manager is at the northern entrance on Madison Avenue. On the second story, besides the Manager's apartments, are nine spacious rooms, study, bath-room, and closets. The third story contains seventeen large and airy sleeping-rooms, besides store-rooms, bath, and closets, while the

THE NEW HOME.

fourth story, in addition to ten separate rooms, contains two hospital rooms and dormitory with adjuncts. The building is heated with steam, and is provided with an elevator.

It is fully capable of accommodating seventy-five men, and the arrangements and management have been perfected after years of intelligent study by the Manager and Directors. Especial attention has been given to sanitary arrangements and to the rendering of the Home attractive and comfortable, in order that the resident's stay may prove agreeable in all respects, as well as beneficial to the body and soul.

The new building, which is free from bond or mortgage cost $125,000 which sum was contributed by the following friends of the work, (the greater portion of these subscriptions are the result of the personal solicitation of Mr. Caleb B. Knevals, one of our Board of Trustees.)

Widow's mite—First contribution toward building the new Home.*			$1.00
(Farabault, Minn.)			
Dodge, William E.	$12,000 00	Dodge, Mrs. Wm. E.	1,000 00
A friend, per Caleb B. Knevals	8,000 00	Gordon, S. T.	625 00
		Kennedy, John S.	500 00
Vanderbilt, C.	7,500 00	Dexter, Henry	500 00
Garrison, C. K.	7,500 00	Hays, Jacob	500 00
Huntington, C. P.	6,000 00	Cauldwell, William A.	500 00
Butterfield, Fred'k	5,300 00	Monroe, Elbert B.	500 00
Stearns, J. Noble	5,000 00	Vermilye, W. R.	500 00
Vanderbilt, William H.	5,000 00	Bliss, Cornelius N.	500 00
Wolfe, Miss Catherine L.	5,000 00	Jesup, Morris K.	500 00
Stuart, Robert L.	5,000 00	Hoyt, Alfred M.	500 00
Talcott, James	5,000 00	Reynolds, C. T.	250 00
Gould, Jay	5,000 00	Willets, Samuel	250 00
Rowland, Thomas F.	4,230 00	Waite, C. B.	250 00
Estate of F. Marquand	3,000 00	Marquand, Henry G.	250 00
Dows, David	2,500 00	Booth, William T.	250 00
Deane, John H.	2,500 00	Inslee, S., Jr.	250 00
Russell, Henry E.	1,600 00	Estate of F. Marquand, per	
Wadsworth, Julius	1,250 00	Alanson Trask	250 00
De Forest, W. H.	1,100 00	Winslow, Edward	200 00
Johnson, John E.	1,000 00	Wheelock, William A.	200 00
Wheelock, William A.	1,000 00	Cotting, Amos.	200 00
Morgan, J. Pierpont	1,000 00	Milliken, S. M.	200 00
McCormick, James	1,000 00	Bennet, Josiah S.	100 00
Lanier, Charles	1,000 00	Havemeyer, F. C.	100 00
Dodge, Norman W	1,000 00	Auchincloss, Hugh	100 00

* From one whose son, as we subsequently learned had, been rescued in the old Home.

Sloan, Samuel	100 00	Harris, Mrs. Rebecca [London.]	50 00
Tiffany, Chas. L	100 00	Redmond, William	25 00
Graham, Malcolm	100 00	Crosby, Rev. Howard, D D.	25 00
Babcock, S. D	100 00	Cash (J. S. & Co,)	25 00
Field, Cyrus W	100 00	Goddard, C	25 00
Stearns, Henry K	100 00	Mrs. B. per J. N. S	20 00
Swan, William H	100 00	Carruthers, George R	10 00
Wales, Salem H	100 00	Becker, Joseph F	10 00
Tucker, John C	100 00	L. M. & Co	5 00
Fogg, William H	100 00	J. W. S.	5 00
Curry, John	100 00	Haffey, John	5 00
DeGraw, W. N. Jr	100 00	Isenmann, John	5 00
Dunn, W. S	100 00	Drescher & Butler	5 00
Terry, John T	100 00	Belden, Rev. Mr	5 00
Daniel, Richard C	100 00	M. L. B	2 00
Rutzler, Enoch	100 00	Davis, G	1 00
Stearns & Curtis	80 00	Belden, Mrs	1 00
Belknap, R. L	75 00	"Grand Blanc"	1 00
Stout, Andrew V	50 00		

From the day of the opening up to the present time, we have heard the song of the new-born soul. God did raise up earnest Christian men, and they, being the true stewards of their Master, were willing and glad to give from the bounty that God had bestowed upon them, so that, as we entered our new Home, we were enabled to enter a Home given to God and paid for by His faithful children. As God has moved at different times upon us by the power of the Holy Spirit, we have advanced in our labor of love, and under His guiding hand our work has been carried forward and we are daily made conscious that it meets with His divine approval. It was in our great weakness that this wonderful work was launched upon us, but our known weakness brought needed strength from Him, and as the years have come and gone He has led us on. My heart is filled with the deepest sense of gratitude to those who have so nobly stood by this work from its infancy, and also with inexpressible feelings of thankfulness to God for raising up such and bringing them to us in our time of need.

A SAINTED MEMORY.

In connection with this work and its great development I find a sweet sense of Christian satisfaction in the testimony borne to its character by our late lamented and esteemed President, Wm. E. Dodge. In the fourth annual report of this institution he said : " We have seen the most wonderful results, and many of the thoroughly regenerated men who are now nobly filling places of trust, might to-day have been filling drunkards' graves but for the timely intervention of 'The Christian Home.' The character of our work is becoming much better known and understood. The facilities it offers for cure, to even men who, commanding high social positions, feel themselves enslaved by the curse of intemperance, and who realize that unless the appetite for intoxicants can be overcome, irretrievable ruin is sure to come, is a marked feature of our system. Such men avail themselves of the influences of our Home without any of the drawbacks of other plans, and find nothing derogatory to their self-esteem, and that no stigma can be attached to their action. The testimony of many who have successfully sought its aid should encourage others to thus find relief."

From the very initiative of our labors, Mr. Dodge proved how deeply he was interested in this Christian endeavor, and from his election to the Presidency to the last moments of his well-spent life he worked with renewed zeal in this portion of the Master's vineyard.

To his generous donations and earnest endeavors we feel to-day that we are greatly indebted for the building we occupy. His presence was always greeted with hearty welcome at our meetings and also his visits to the Home. From the first I can say that I felt we had in him a most noble Christian friend, a kind-hearted and generous supporter, so like his Master in all his doings. When Mr. Dodge was called from our midst we all felt that the Home had sustained a severe loss, not alone of a sincere friend but a wise and prudent counsellor.

At one of our meetings in November, 1882 (the last one he ever attended in the Home), after giving an account of his own conversion, he further said : "I have listened to each of your testimonies, and it has been a source of great comfort to me. I have noticed all through the meeting that a feeling of gratitude fills your hearts. As you have told of the love of God manifested in your conversion, so also have you expressed gratitude to the founders and benefactors of this Home. I for one would say I have been fully repaid for all that it has been my privilege to do in helping to establish and build this Home while listening to your prayers and testimonies this evening."

A LIFE OF BROADCAST BLESSINGS.

At the close of this meeting Mr. Dodge, on learning of Mr. Jay Gould's donation of $1,000 on condition of the remaining debt of $5,000 being paid, said to Mr. Knevals, "I will also give you $1,000 toward this." This was his last contribution to the Home.

Distinguished among New York's bright array of philanthropists was our late friend, Wm. E. Dodge, of whose generous, varied, and quiet benefactions the half has never been told. In public and private acts, his was the warm loving charity that did indeed bless him who gave, as well as him who received, that shrank from publicity, and often gave in secret, content to diffuse its fragrance and bestow its sweetness, without thought of gratitude or praise. From the founding of "The Christian Home," in which he was a prime mover, to his lamented death, Mr. Dodge manifested the deepest interest in its growth and success.

His zeal to promote the welfare of the Home never flagged ; his contributions to its establishment and support were worthy of his great heart, Christian character, ample fortune, and of a great philanthropic work. To his personal influence in behalf of the "Home," and active interest in its management, is due a very large measure of its success. His earnest cooperation at the very inception of the enterprise, and continued sympathy with the Resident Manager, in his arduous and difficult position, were keenly appreciated and will ever be gratefully remembered.

Words, however eloquent, can pay but a feeble tribute to the memory of one who scattered his blessings broadcast, to bloom forever fresh and sweet. When such a life goes out from us it leaves, like the setting sun, a great glory in the path of its departure, with this difference, that the glory never fades.

At a meeting of the Board of Directors of The New York Christian Home for Intemperate Men, held on the 13th of February, 1883, the following preamble and resolutions were unanimously adopted :

Whereas, It has pleased Almighty God to take from our midst our late beloved Associate and esteemed President, the late Wm. E. Dodge, it is fitting and timely that this Board should place upon record their tribute of respect to his memory ; be it therefore

Resolved, That in the death of Wm. E. Dodge The New York Christian Home has lost a devoted and sincere friend, a wise and prudent counsellor, and a generous supporter. His unblemished Christian character, his zeal in doing good, his sympathy with distressed humanity, his almost unbounded generosity toward his fellow-man, and his unvaried courtesy entitled him to our profound regard and esteem.

Resolved, That we tender to the family of our deceased friend and co-worker our heartfelt sympathy in this hour of their affliction.

Resolved, That these resolutions be placed upon the records, attested by the signature of the Vice-President.

J. NOBLE STEARNS, *Vice-President*.

At the gathering for family prayers in the chapel of the Home on the morning of February 10th, 1883, the following preamble and resolutions with respect to the death of Wm. E. Dodge were offered and adopted by the resident members:

Whereas, Since we last met for morning prayers, God in His providence has called from us our beloved President, friend, brother and benefactor, William E. Dodge, after a long and useful life. A faithful servant of his Master, he was always doing God's bidding, and, like his Master, his heart was always touched on hearing of the wants, the sorrows and the troubles of those about him. We must believe that his work in this life was finished, and that he has been welcomed into the loving presence of his Lord. Oh, how comforting is this thought to his friends. "So the Lord led him, and there were no strange gods with him."

Whereas, Feeling as we do the great and irreparable loss which all connected with the "Home" must ever experience, not only in the death of its benefactor, but in the absence of the kind and generous friend and adviser, we desire to express, as far as possible in these few words, our deep sorrow. Therefore, be it

Resolved, That we extend to the widow and family, in their hour of bereavement, our most heartfelt sympathy.

Resolved, That a copy of this preamble and resolutions be sent to the family. On behalf of the members now residing at the "Home."

Signed, CHAS. A. BUNTING.

Passages of Scripture that Mr. Wm. E. Dodge would have read in the morning and evening of the day of his death, from his selection of Scripture for every day in the year :

And I heard a voice from heaven, saying unto me, Write, Blessed are the dead which die in the Lord from henceforth: Yea, saith the Spirit, that they may rest from their labors; and their works do follow them.—Rev. xiv. 13.

Whatsoever thy hand findeth to do, do it with thy might; for there is no work, nor device, nor knowledge, nor wisdom, in the grave, whither thou goest.—Eccl. ix. 10.

For I am now ready to be offered, and the time of my departure is at hand. I have fought a good fight. I have finished my course, I have kept the faith: Henceforth there is laid up for me a crown of righteousness, which the Lord, the righteous Judge, shall give me at that day: and not to me only, but unto all them also that love his appearing.—2 Tim. iv. 6, 7, 8.

There remaineth therefore a rest to the people of God. For he that is entered into his rest, he also hath ceased from his own works, as God did from his. Let us labor therefore to enter into that rest, lest any man fall after the same example of unbelief.—Heb. iv., 9, 10. 11.

Thy sun shall no more go down, neither shall thy moon withdraw itself: for the Lord shall be thine everlasting light, and the days of thy mourning shall be ended.—Isaiah lx. 20.

He will swallow up death in victory: and the Lord God will wipe away tears from off all faces; and the rebuke of his people shall he take away from off all the earth: for the Lord hath spoken it.—Isaiah xxv. 8.

And I said unto him, Sir, thou knowest. And he said unto me, These are they which came out of great tribulation, and have washed their robes, and made them white in the blood of the Lamb.—Rev. vii. 14.

They shall hunger no more, neither thirst any more: neither shall the sun light on them, nor any heat. For the Lamb which is in the midst of the throne shall feed them, and shall lead them unto living fountains of waters: and God shall wipe away all tears from their eyes.—Rev. vii. 16, 17.

STEADFAST AND TIRELESS WORKERS.

Befitting as it was to place the wreath of immortelles upon the brow of our dear departed President, we feel it our bounden duty to thank and praise God for the directing power He gave us in selecting his son, Rev. D. Stuart Dodge, upon whom the father's mantle has fallen, and who is now filling the place as the third President of the Home. From his first occupancy of the office, Rev. Mr. Dodge has filled the position to our utmost anticipations. He has nobly responded to every call, and has been most earnest in every act of the past years of his incumbency.

Mr. James Talcott was our Treasurer for five years, filling the office with marked fidelity, ever ready to help us in times when we were not prepared to meet our immediate indebtedness. He was a shrewd and calm financier, and an earnest advocate of economy in all that related to the expenses of the Home. Mr. Talcott's resignation as Treasurer in 1881 brought to us Mr. Frederick A. Booth, a sincere and consistent Christian, and his labor of love we feel has been appreciated by every member of the Board of Directors.

Mr. Caleb B. Knevals was one of the original founders, and has filled the office of Chairman of the Executive Committee continuously, besides acting as Trustee under the reincorporation of the society. His active and untiring zeal has been invaluable, not alone in securing contributions for the support of the Home from year to year, but to him are we largely indebted for the money raised to erect and complete our present edifice.

Our freedom from debt and ability to meet our daily demands are but another evidence of Mr. Knevals' continued and laborious efforts. When I say that I deeply appreciate and honor him for his noble support and advice, especially in times of discouragements and perplexity, always ready to give words of cheer and comfort, I but feebly express the sentiments of my own heart; and we have been signally blessed in his self-sacrificing devotion to our Board of Directors all through these ten years, deeming it a pleasure to give to this work a great deal of time in planning and devising the ways and means of carrying it forward.

We feel here it is but just that special mention be made of the marked interest shown in this work by Alonzo S. Ball, M.D. From its inception, and before his name was added to our list of Directors, since then and down to the present day he has been in constant attendance at our Saturday evening meetings. To him we are deeply indebted for no little excellent counsel, and his zeal in encouraging us is a constant source of help and consolation. The age, experience, deep religious culture and character of Dr. Ball imbue his opinions with weight and meaning. As in the case of Dr. H. C. Houghton, who declared at one of our Saturday meetings that his whole Christian life, character, strength and love had been moulded and grounded through the influence of "the Christian Home" during the seven years that he had been a regular and frequent visitor at the Saturday evening meetings, adding that the men before him were actual Christian teachers and moulders in the world, so Dr. Ball, while earnestly endorsing the language of Dr. Houghton, said that his faith in the unlimited power of God, especially His saving and keeping power had been mightily strengthened by the many years' experience which he had gained at the religious meetings of "the Christian Home."

Since the election of our honored and esteemed friend, Mr. John Falconer, on the Board of Directors in 1881, he has been of singular utility to this institution, being one of the most valuable acquisitions made to our ranks of active and unselfish friends. His constant attendance and prayerful spirit at our public weekly meetings are sources of inspiration and encouragement to all. A devoted Christian, he is no less a sympathetic yet prudent adviser. His interest never flags, so filled is he with the feeling he so frequently expresses that the " work of The New York Christian Home is wonderful and above comparison with any other form of Christian endeavor in the same direction," and from out the abundance of his heart he pleads for its needier inmates among his friends, inducing them from time to time to send welcome contributions of money and equally welcome clothing, etc., for the support and use of those who so urgently require so full a measure of Christian liberality. For myself personally I feel that no words can express how gratefully I recall his attentions during the weary months of sickness I passed through. It was then that he poured the richest fruits of his soul by my bedside in prayers to the throne of grace for the assuagement of my pain and my recovery. New lights and freshened hopes he brought to my chastened spirit, and as he dealt with me in my hours of trouble so may his kindness be remembered by Him who has promised reward for even a cup of water given in His name.

OFFICERS AND DIRECTORS.

The following is a complete list of the Officers and Directors as they have served from year to year, many of whom, it will be noticed, have acted from its commencement to the present time:

OFFICERS 1877, 1878, '79, and '80.

Wm. T. Booth, elected President 1877, resigned 1880. Wm. E. Dodge, elected Vice-President 1877, elected President 1880, died February 9, 1883. Arthur W. Parsons, elected Secretary 1877, resigned in 1880, died in 1884. James Talcott, elected Treasurer in 1877, resigned 1881. Charles A. Bunting, Resident Director, elected 1877, acting.

THE NEW HOME.

CHANGE OF OFFICERS IN 1880.

Wm. E. Dodge, elected President; J. Noble Stearns, elected Vice-President; M. C. D. Borden, elected Secretary.

SUBSEQUENT CHANGES.

Rev. D. Stuart Dodge, elected President 1883; Frederick A. Booth, elected Treasurer in 1881; Henry C. Houghton. M.D., elected Secretary 1883.

TRUSTEES.

J. Pierpont Morgan, C. N. Bliss, C. Lanier, Bowles Colgate, G. H. Andrews (resigned), and Caleb B. Knevals.

BOARD OF VISITORS.

Rev. W. M. Taylor, D.D., Rev. R. S. Macarthur, D.D., Rev. W. F. Watkins, D.D., Rev. R. R. Booth, D.D., Rev. A. D. Vail, D.D.

BOARD OF DIRECTORS

From 1877 to 1887, acting at various periods.

William T. Booth, elected 1877, acting.
Wm. E. Dodge, elected 1877, died Feb. 9, 1883.
Caleb B. Knevals, elected 1877, acting.
Elbert B. Monroe, elected 1877, resigned 1880.
J. Talcott, elected 1877, resigned 1884.
J. Milton Cornell, elected 1877, resigned 1879.
John N. Stearns, elected 1877, acting.
R. R. McBurney, elected 1877, resigned 1886.
Arthur W. Parsons, elected 1877, resigned 1881.
Samuel C. Burdick, elected 1877, resigned 1879.
Edmund Penfold, elected 1877, resigned 1879.
D. Stuart Dodge, elected 1877, acting.
Willard Parker, M.D., elected 1877, resigned 1880.
W. R. Vermilye, elected 1877, acting.
Chas. A. Bunting, elected 1877, acting.
H. Dexter, elected 1878, resigned 1887.
N. W. Dodge, elected 1878, acting.
A. C. Armstrong, elected 1878, acting.
Richard A. Storrs, elected 1878, acting.
S. Sheldon, elected 1878, resigned 1880.

J. Edgar Johnson, elected 1878, resigned 1886.
M. C. D. Borden, elected 1878, resigned 1885.
Francis A. Palmer, elected 1878, resigned 1880.
A. V. Stout, elected 1878, died 1883.
B. Colgate, elected 1878, resigned 1880.
Wm. A. Cauldwell, elected 1878, resigned 1881.
C. Vanderbilt, elected 1878, acting.
Henry E. Russell, elected 1879, resigned 1887.
Thomas F. Rowland, elected 1880, resigned 1887.
W. H. Jackson, elected 1880, resigned.
J. H. Dean, elected 1880, resigned 1882.
John Falconer, elected 1881, acting.
Fred'k A. Booth, elected 1881, acting.
W. M. Isaacs, elect'd 1882, resigned 1883.
A. S. Ball, M.D., elected 1884, acting.
Clinton B. Fisk, elected 1884, acting.
Dr. H. C. Houghton, elect'd 1884, act'g.
Titus B. Meigs, elected 1885, acting.
James H. Dunham, elected 1885, acting.
Dr. Wm. E. Dunn, elected 1885, acting.
James H. Seymour, elected 1887, acting.
D. H. Bates, elected 1887, acting.
Samuel W. Bowne, elected 1887, acting.

FRUITS OF THE WORK.

Immediately upon opening the Home, ten years ago, we were supplied with ample material to commence our work. The first who came to us was a business man well known in the city of New York, formerly an importer. We received him kindly and administered to his outward wants, clothing was furnished from head to foot and his famished body supplied with necessary food. Let me say that this man was soon brought to a knowledge of the truth as it is in Christ.

After remaining with us a few weeks a position was secured for him, as manager of a concern in a neighboring state. For ten years he had not held a position of any kind, being constantly under the influence of liquor and his family was separated from him. Soon after his arrival in his new home he connected himself with a Christian church and is now in Chicago a faithful worker in the Gospel temperance cause.

Shortly after this as I was sitting at my window looking into the street, saw a man dressed literally in rags. He came to the door and was admitted. I soon learned his errand. He told me his sad story, and in naming over his friends he mentioned two persons whom I knew. I found he was a man of education and had stood high in his profession as a lawyer, having been a student and reader of law with Hon. Edwin M. Stanton. He soon gave his heart to God and is now living a devoted Christian life. I received a letter from him soon after his arrival home in Penn. and these are his words: "I have just arrived home after an absence of four years. I have told sister all. She is supremely happy and so am I, thank God and the dear Christian Home."

CONVERTING A SCOFFER.

Another instance. While holding a meeting on a Sunday afternoon in the Gospel Tent, I mentioned in the course of my remarks something in relation to this Home. A widowed mother came to me and related the sad tale in relation to her only son. She informed me as to his unbelief and said "I have no hopes of him." After advising her what course to pursue she said "I will try and get him to go to the Home."

A whole week passed before he came. He remained in the Home over a week before I could approach him in conversation. This case seemed to stagger me. He was a professed infidel, a scoffer, but I prayed to God and He answered my prayer. I felt fully assured that this man would be saved. For the first time I called him to my room, for the purpose of conversation and prayer. I stated to him my knowledge of his unbelief. I said "I know you pretend not to believe in a God, the Bible nor have you any confidence in the Christian religion. I want to pray with you and I wish to make a proposition to you. It is this, if you can bow with me without prejudice and with an unbiassed mind, and in a sincere manner say to me that if I can prove to you that there is a God in heaven and that the Bible is true and that this is the way of salvation, as a candid man that you are ready to accept it? If it is not proven to you before you rise from your knees, that there is a God in heaven and that He has power on earth to forgive sins and that the Bible is true, then I will not speak to you again on this subject, and you can lay the Bible one side." He accepted the proposition and we bowed in prayer and while on his knees he prayed "God be merciful to me a sinner." From that moment he became a changed man. God heard his prayer.

Another was a policeman, who for a number of years was on the "Broadway Squad." He like the first, came to me in a destitute and wretched condition, having been separated from his family for many years. He gave his heart to God, was united with his family and by his godly example and Christian life, his wife and daughter were soon led to follow. They united with a Methodist church in this city, the pastor of which told me since, that this whole family were considered a valuable acquisition to the church.

"THE GOSPEL TEMPERANCE SALESMAN."

A gentleman was brought here in the summer of 1878, by his nephew, in a most dilapidated condition. We furnished him with clothing, and, as soon as practicable afterward, I entered into conversation with him. I found him to be a man of education, thoroughly acquainted with business life, and at one time had been at the head of a firm. He had been employed as a salesman in this city, but because of his habits was dismissed. He became a willing convert to the religion of Jesus Christ, and gave us unmistakable proofs of his conversion up to the time of his leaving the Home. He immediately obtained a position. As soon as he received his first month's pay, which was only $30. he came and paid up two week's board, and the next month paid up the balance. (I am happy to say that this is not an isolated case, many of those so taken in, returning and paying their board.) He was quickly sought after by business houses, from one of which he obtained a lucrative position, and at this date I learn from the best authority, that he is giving entire satisfaction as a commercial traveller. He is in regular correspondence with the Home, and never visits this city without coming to our meetings in company with his wife and giving his testimony. He is known over the whole route which he travels as "The Gospel Temperance Salesman." In his prosperity, he does not forget the Christian Home, and is most generous in his donations, and never fails to mention his spiritual birthplace. As we look upon him we can say, in the language of Isaiah, "Who hath wrought and done it? I the Lord, the first, and with the last; I am he."

A devoted Christian young lady who for many years visited Blackwell's Island every Tuesday, and who is a frequent visitor at our Saturday evening meetings, came to me one day and said: "I have found a very interesting case on the Island. He is the son of a clergyman. Won't you take him into the Home?" He was received. Full of scepticism and unbelief he came to us. I had frequent conversations with him, and so far as I could judge, all without avail. I found it was useless to exhort, and told him I should not again urge upon him the claims of the Gospel of Jesus Christ until he came to me with all sincerity of heart, feeling the need of a Saviour. These must have been the words that had the desired effect, although apparently he was determined not to yield. During the three days following this interview he was constantly before me, but not a word did I say to him personally upon the subject. At last he said to me, "Have you given me up?" I recalled to his mind the last remarks I had made to him, when he exclaimed in great earnestness "I do feel the need of Christ. Will you pray with me?" I took him to my room and he then and there consecrated himself to Christ. His life since that day has proved him to be a true follower of Jesus. He now holds a responsible position, is an active member in the church, is engaged in mission work, and is loved and respected by all Christians in the city where he resides. He is a noble monument of God's power to save. Listen to a few words from his father: "Our family, (his brother and sisters,) are surprised and delighted at the marked and marvellous change which appears more and more clearly in each succeeding epistle that comes from their elder brother. His oldest sister just remarked to her mother, 'of all the wonderful things I ever heard in relation to such matters, the conversion of —— is the most wonderful' and all the rest are disposed to say Amen."

A noble young man occupying a public position in the city was directed here by a kind Christian gentleman. You cannot conceive how low he had fallen. For weeks sleeping in wagons, hallways, lumber-yards, and parks. His scanty rags did not cover his body. He was admitted, and at once accepted Christ. He is a worthy Christian, and is rapidly gaining the respect and esteem of those who once knew him.

REV. DR. KING SPEAKS.

I might give many other instances, but will restrict myself to one only as related in one of our Saturday evening meetings by Rev. Dr. James M. King, of the Park Avenue M. E. Church, who commenced his remarks by saying that the great feature of the Home is the *success* of the work, and this fact is always observed here, that *if God saves a man, that person knows it, and does not need any affidavit to prove it. God tells it.* He said that he did not know of any work in the world which is more the sole fruit of faith than this temperance work in the Home. It is this which saves men. Ever since he had been a Christian he had large sympathy for the drunkard, and the drunkard does not need any other pardon than is needed by any other sinner, and very often the man who turns away from the drunkard with mistrust or scorn is far more in need of a Saviour than the poor intemperate man. Dr. King said he knew a great many trophies of grace which came from this Home, but the brightest to him was the following: "Three years ago I went into a miserable home in this city to see a poor woman who was suffering from her husband's dissipation. She was once beautiful, but the anxious look told plainly what she was suffering. Two or three little children were there, so scantily dressed that they could not go out of doors ; all shivering ; but that dear mother never complained, but worked on, and ever prayed. There was but little furniture in the house, and a more dreary picture I have seldom seen. The husband, who was a reeling drunkard in the streets of New York, was once as fine a fellow as you meet. He was the son of a prominent clergyman of this city, but friends and kinsfolk had done all they could for the boy, and had almost given him up as lost. But a few of us got together, and resolved to make one more effort to save him, and we sent him to this Home ; and you all know what he learned, for you have the same teaching.

"The other day a fine looking man called at my house, dressed well, fresh looking, and manly looking in his actions. He wanted to know if I would go over and preach at *our* church next Sunday. I knew him and asked him what he meant by 'our church.' He said he lived in New Jersey, and was a communicant of a church there, and in fact was one of its trustees. I grasped him by the hand and said, God be praised ! It was that same poor drunkard, now a *man*. I went, and found a beautiful little home without a dollar of debt on it, his dear wife and children as happy as they could be ; the husband back in his former business, and just admitted as a partner, and looked upon in that village church as a leader and wise counsellor. This is what the Home has done for one man, and if it has never accomplished more, those godly men and women who gave their money to build this place have saved a noble human soul from eternal death.

OUTSIDERS BROUGHT TO JESUS.

The following are a few of the many instances of conversion of outsiders, not members of the Home, while attending the meetings. The first was the wife of one of our inmates, who was a constant attendant while her husband was residing here. The Holy Spirit gained admittance to her heart, and here she found peace in believing on Jesus. From this her daughter became interested, and she too became a Christian, and they were united to the Methodist Church. They are now living, active Christians, and their example is felt in the neighborhood where they reside.

A mother was induced to bring her son to this Home, and in conversation with her I found that she was not a Christian. The plan of salvation was made known to her. She accepted Jesus as her Saviour, and I believe she is now a good Christian.

One afternoon a lady called to see her husband. He was attending at the time one of our afternoon prayer-meetings. In my remarks to her I said, "Now your husband has become a Christian, I do hope you will encourage him by your godly life.' She exclaimed, "Is my husband a Christian? I want to be a Christian too!" As he came from the meeting into her presence the first question she asked was, "Are you a Christian?" calling him by name. He answered that he believed he was. We all knelt in prayer immediately, and it was my privilege to see her rise from her knees a converted woman. They have walked hand in hand, in their Christian way, from that day to this.

A father came here, and as he told me about his son my heart was moved toward that family. That son became an inmate of this Home, and soon a Christian man. His father saw what a change the religion of Christ had wrought in his son, but although a very honest, upright man, had never known what the new birth meant. He gave his heart to Jesus, and is now a happy Christian man. The wife of the son also came to visit the husband, and in conversation with her I found she too had lived a stranger to the great truths of the Gospel of Jesus. It was my privilege to kneel in prayer with her and hear her confess Christ before leaving the room. They too are living active Christian lives.

A wife came here to be prayed for. The plan of salvation was explained to her. She was ready to accept Christ and believe on His name. Her husband had been an inmate, and they are now going along on their new road heavenward.

A fine looking gentleman called here one day to make arrangements for his nephew. From his appearance I judged he was a Christian man. I explained to him the nature of our work in detail, to which he paid strict attention. The next day he came with his nephew, and then we had another conversation. He called the third time, and in my interview with him he informed me that he had been so impressed in listening to me that he had been induced to close his bar, which he had been running upward of thirty years, in connection with his restaurant, located on one of our principal thoroughfares. I found that he was under deep conviction, and I wanted then to pray with him, but he declined and made an appointment for the next day. He kept his promise. We were together two hours or more, and it was my happy privilege to lead him to Christ. I see him frequently, and his expression always is, "This life is worth living. I am now a happy man."

A man came here, sent by one of our leading physicians. He was in a pitiable condition, but in earnest as to his soul's conversion. As soon as he became a Christian it was his great desire that his wife might become a Christian also. He asked prayers for her in our meeting. She called to see him, and I was introduced to her. In her conversation she said, "My mind has been so troubled for the past week that I cannot sleep. I do not know what is the matter with me; but one thing I have made up my mind to do, and that is as soon as you come home (turning to her husband) to become a Christian." He said, "We have been praying for you for a week," and asked me if I would then talk and pray with her. I did so, and it was my privilege to see her leave the room a Christian woman.

At one of our late meetings a testimony like the following was given by a bright, intelligent man, who had prepared himself by arduous study to preach the Gospel of Christ. Through the advice of a physician, he was induced to use morphine to alleviate pain. Sufficient to say, that for eight long years he had been bound by his appetite for the drug, and was on the verge of despair when he entered the Home. Kind words and kind treatment won him first. He was told if he would place himself under our care he would be treated kindly, and that he would have to endure but little suffering, as it was the Great Physician to whom we would apply in his behalf. He was assured that the household would offer prayer for him constantly. God saved him, and now he is a bright and shining light.

PHYSICALLY AND SPIRITUALLY TRANSFORMED.

Among our achievements is the not infrequent transformation of the "bummer beat" into the Christian gentleman. Precautions are taken against impostors, men who will profess anything for the sake of shelter, food, and needed clothing, but some such fellows do manage to creep into better society than they deserve to enjoy. They are soon detected after admission, but their presence is ordinarily winked at as long as they behave themselves. Their reformation is not despaired of, and nothing is said or done which can inform them that their deception is seen through. They are treated exactly like the rest of the inmates, with the same courtesy, forbearance, and kindness, and the same reforming influences are brought to bear upon them. The result is, in not a few cases, that these "meanest of mankind" become changed characters. One's business or professional training modifies the views taken of any subject; hence my interest in the physical as well as the spiritual features of the work. A moment's review brings to my mind a score of men who were physical wrecks, entering the Home in such conditions that an exact and complete statement of it would be simply revolting. These men were promptly and perfectly restored. Cleanliness, quiet, good nourishing food are the means used on entering, and in special cases minimum doses of sedative medicines are administered to control delirium; but it must be distinctly understood and recognized that this is exceptional and not the object or method of our work.

Alcoholic patients furnish not a few examples of those who had passed beyond a point where ordinary spirits had any effect, and the use of raw alcohol, chloroform, etc., had been resorted to. One patient, a practicing physician, who came to us confessed that the following drugs and stimulents had been his regular rations for months past, viz.: Nux vomica, ipecac, ammonium, bismuth, blue mass, calomel, camphor, triplex valerian, morphine assafœtida, raw alcohol, and whiskey. This man was actually known to have made inquiries in the drug stores as to whether anything new had been discovered in the way of nerve disturbers. He left this Home a reclaimed man, and is now much respected and in the enjoyment of his practice, which for the most part he has recovered. His testimonies can be frequently heard in our Saturday night meetings.

DR. HOUGHTON'S VIEWS

Dr. Henry C. Houghton speaks on this point as follows :

"The physical restoration following upon the spiritual birth is more surprising, a marvel not easily solved. The reflex influence of a healthy mind upon a diseased body we understand ; but how the physical wreck of alcoholism can be restored so suddenly, or how in exceptional cases kept from death, is not clear even after years of observation. The transformation that takes place in the bodies of these men is a greater surprise than the one in the spiritual nature ; we expect a greater or less degree of functional or organic disease to follow a long history of alcoholism, but in the greater proportion of cases the recovery is rapid and permanent. In exceptional cases the reflex influence of the soul life has been the only reasonable explanation of the maintenance of physical existence.

"The character of the religious life is also exceptional. From a depth of sin and degradation, they advance rapidly to a degree of assurance and stability that is not usually noticed till one has attained to some degree of maturity in the Christian life. Some, it is true, relapse, but the great majority stand, manifesting an earnestness of purpose, a clearness of perception of divine truth, a love of God and their fellow men, that those who have been Christians for years may emulate.

"Redeemed men go from the Home to work for the Master as none others can. Ten clergymen are now doing an effective work like Peter, loving more intensely, having been forgiven much. Many of the lay members are nightly and on the Sabbath active workers in the various missions of the city. 'The last shall be first;' God has taken those whom our faith had not dared to claim, and in these days placed them in advanced positions, both as regards personal experience and as missionaries to the lost.

"To such a work we solicit your attention. To such a work we ask your support, by prayer, by your donations, by your presence in the Home, thus encouraging others and being blest yourself, as you see and share the goodness of God."

FACTS FOR CHRISTIAN THINKERS.

The remarks of our respected Secretary just quoted are no less replete with truth than with the wisdom which is born of charity and developed by years of experience in Christian labor. I know that it has not been thought advisable, even by the kindest hearted men and women, to sympathize with the drunkard ; but my dear friend, whoever you may be, let me tell you that there is not a class of men on the face of the earth who need your sympathy more.

That we gather the objects of our care exclusively from the depraved and vicious classes is a false idea, and we seek to remove it from the popular mind. The privileges of the Home will never be refused to any sincerely desirous of freeing themselves from the appetite and bondage of intemperance, yet for the most part our work has been among men of culture and former position who have fallen into this life.

Among those who have gained admission to the Home are many who have been born and bred to high social positions in the world. They have been graduated in colleges and have become clergymen, lawyers, physicians, and merchants, but the drink habit had undone them. If they are ever to recover and become true men again, it must be by the restoration of the will, and Christ, it is believed, is the only Physician who can restore it.

The fine and family appearance of all the house arrangements greatly aid in the work among this class of inmates. All prejudices against asylums are removed as soon as they enter our cheerful, well furnished, attractive house. Their uniform restoration to health has been most marked. We dare but advocate one principle—total abstinence.

FILL THE HOME.

What we desire, what multitudes of suffering, struggling, degraded despairing men desire, is to have this Home filled. Its ample accommodations should all be in constant use. This can readily be done if friends of temperance, if the friends of these needy men from every rank of life, would supply the means. The Home does not receive aid from the Excise Board. Its friends question the propriety of endeavoring to cure drunkards by licensing drunkenness. Nor does it obtain help from City or State funds. It is deemed inexpedient, if not harmful, to the best interests of the community for such an institution to apply for appropriations from the public treasury. The work is carried on by the free gifts of all who feel the sore need of every agency to save intemperate men, while it especially appeals to those who wish to encourage in such efforts the wider use of direct religious influences.

No one's heart can fail to be moved which knows the private history of these reclaimed men. Among them we find those who have been drunkards for five, ten, twenty, thirty years, and upwards. As our mind's eye glances over the lists we think of the many who have been arrested in their downward career that are now honorable members in society and filling prominent places in Christian churches. Many of these have gone into the Gospel field, laboring successfully in leading poor lost men to Him who saves to the uttermost.

A NATIONAL INFLUENCE.

We have graduates from this Home preaching this Gospel Temperance, either as settled pastors (of whom, by the way, we have many) or evangelists or laymen, representatives in all these States I name : Alabama, Connecticut, California, Delaware, Illinois, Indiana, Kansas, Maine, Massachusetts, Maryland, Michigan, Missouri, Minnesota, New Hampshire, New York, New Jersey, Nebraska, North Carolina, Ohio, Pennsylvania, Rhode Island, Vermont, Virginia, Wisconsin, and Washington, D. C.; besides in England, Ireland, Scotland, Canada, and Australia.

A prominent gentleman was in attendance at one of our Saturday evening meetings not long since, and made this remark : " Wherever I go

I find representatives of this Home. In thirty-five missions which I have been privileged to visit within the last three months, I have found men who said that the Christian Home was their spiritual birthplace." So you see the influence of this Home is felt all over our land. Oh, that Christian ministers might spread the good news, and religious journals do the same, that there is one place where the salvation of the drunkard is relied upon through the agency of the Holy Spirit. I would also state that if men were encouraged by their friends to spread this good news after leaving the Home, by testifying to the saving power of God's grace, much more might be accomplished. The friends of the saved ones have that false pride still clinging to them, and they rather discourage anything of the kind for fear of the disgrace that might come from such testimony. Another thing I would like to say to the friends of those who have been saved in this Home, is by all means encourage them to visit the Home, and attend our regular Tuesday or Saturday evening meetings. Come along with them, and rest assured you will always be kindly welcomed. Also to the friends of the former members let me say this : Encourage the erection of the family altar. Also encourage your friends to select a home in some live Christian church, where they will be missed and inquired after in case of their absence from the regular stated services.

THE BIBLE OUR TEXT-BOOK.

Frequently the question is asked, " What are the amusements, and what do these men do ?" Our reply is that as every one who enters this Home comes with the desire (or professes to) of changing the tenor of his life, it is readily seen that the few weeks which are spent here cannot be put to a better use than in learning how to live the new life that all are supposed to enter upon. As our only text-book is the Bible, which is comparatively a new book to the many who come to us, we find our time profitably employed in its study.

Henry Ward Beecher expressed our theory pretty thoroughly when he said : " By the power of the Holy Spirit men are transformed, inspired, and brought into a state in which it is not mockery when they are called sons of God. So that it is the avowed opening of this new kingdom of influences, it is the direct inspiration of the human soul by divine contact, that constitutes the peculiar operative element of the New Testament. The truth that the Holy Spirit of God acting upon the human soul develops it in all those qualities which are farthest from the animal and nearest to God, is that one truth of the New Testament which inspires the most activity, the most rational hope, and the most practical development of Christian efficiency. That power of God (the Holy Spirit) acting in the human soul, kindles the imagination, fires the reason, creates a moral enthusiasm, and gives to the latent or undeveloped resources in man power by which he becomes a son of God in disclosure as he was before potential in his undeveloped condition." Let me say, we have seen no reason, since the beginning of this work in the old Home in 78th Street, to change our manner of dealing with the drunkard. After he is received, he is cared for by faithful, efficient Christian nurses. If the man is physically diseased, and needs medical attendance, medical aid is at once summoned. Medicines are required for sick drunkards as for any other class of people who have suffered from exposure and from excessive use of stimulants. As soon as the patient is in condition, he is removed from the hospital to his room. No medicine is given in this Home for the purpose of destroying the appetite or for weaning the man from his cups. Not a drop of liquor of any kind is given to "taper off," as is a common custom.

TOBACCO A SNARE.

The use of tobacco in any form is, as reference to our rules will show, strictly forbidden, and we strongly advise all our members to abandon the habit forever, to cast it out with their other sins, because we know it is a great stumbling-block to the young Christian, and in too many instances has proved to be the besetting sin. Could we forever banish tobacco, the twin brother of rum, I feel confident that there would be no cause to fear the future of most of the men who have been under the influences of this Home. Tobacco is the rock on which many converts make shipwreck of their faith. Out of the 1878 men who have professed to be saved in this Home, not one, to our knowledge, has returned to the old vice of drink who has abandoned his tobacco. Do you not see, then, how our hearts are pained when we know of Christian leaders who are addicted to this habit of tobacco using? Can any one pretending to be a Christian teacher and leader successfully answer these two questions: If tobacco is a poison of a deadly nature, which, used by chewing, snuffing, or smoking injures the mouth and throat, the voice, stomach, and digestive organs produces debility, failure of appetite, indigestion, constipation of the bowels, injures the complexion, the lungs, the heart, poisons the blood, destroys the brain and nerve power, impairs the memory, deadens the sensibilities, creates a craving for stimulants, and makes one filthy even to nastiness—can a conscientious Christian man or woman use it with impunity, asking God to sanctify it to the benefit of his or her body? If not, how before God can we dare to ask Him to purify or make us clean when we are constantly befouling ourselves with that weed? We have invariably found it to be the case that a return to the tobacco habit involved a relapse to the drinking habit, and I never feel secure as to a member until he willingly resigns tobacco forever.

After the man leaves the hospital (which is almost invariably on the second or third day after entering the Home), he at once attends all our religious exercises, and when he has had the privilege of attending either the Tuesday or Saturday night meetings, he is taken to the room of Mr. Pulis, my assistant, or to my own private room, where the plan of salvation is made known in all simplicity. A great majority accept Christ then and there. After this interview, and their profession of faith, it has been our custom to expect these men to openly confess Christ with their mouth (Rom. 10:9) in our Tuesday evening meetings. By this method we are able to see just where they stand, and the progress they are making in the new life.

TELLING TESTIMONIES.

In connection with this portion of the fruit so plentifully growing in this interesting section of the Lord's vineyard, I deem it well to subjoin the following testimonies, taken as they fell at one of our Saturday night meetings from the lips of men who had left the Home at various periods of time, but all concurring in ascribing their miraculous change of heart and life to the power of Jesus:

TELLING TESTIMONIES. 47

Mr. W. (five years true to Jesus): "When I first came to this Home I was met with such a greeting that I can never forget. How different the treatment to what I used to receive from my fellow beings. My heart continues to go out in praise that He established such a place for intemperate men. Had it not been for this Home God only knows where I would have been. As I reflect upon my condition then I fail to find words fit to express my praise to my Saviour. God knows my heart, and that it is full. He knows whether or not I am honest. It is my determination to make heaven my home. I have tried the world long enough; now I have come to the Saviour."

Mr. S. (over seven years steadfast): "I feel it a great privilege to participate in this meeting. I have for many years been a slave of intemperance. I have no words at my disposal to convey even an imperfect idea of what I have suffered, mentally and physically, through the cursed sin of intemperance. It drew me down, down, until I was powerless to release myself from its relentless grasp. When I was induced to come here I was a skeptic. I am much indebted to Mr. Hayes for his uniform kindness. He suggested that I should go to the hospital for a time, and I came out very much improved in health, but spiritually still a very sick man. I came back to the Home still doubting. I heard the testimonies of different men, and thought they were doing it to please somebody else. But finally I noticed that the earnestness and sincerity of what they said stamped it with truth, and I reasoned with myself : ' If these men's statements be true, and they were all as I am, and have found salvation, surely there is hope for me.' I went on from day to day (and I am heartily glad to see our Manager in his accustomed place to-night, for to his teachings and his able assistant I am indebted to a very great extent for the marvellous change wrought in my life), and I prayed earnestly that God would take the appetite for drink from me; and He in His mercy answered my prayer, and I stand here therefore to testify to His power and willingness to save and to keep."

Mr. C. (faithful for three years): "I journeyed fifteen hundred miles to this city from St. Louis for the purpose of entering the Christian Home, and I wish to-night to testify to all that it has done for me through the grace and mercy of God. Before I came to this blessed place I had been in nine institutions for treatment, but none of them were of any benefit to me. When I left here, after a stay of six months, I was a different man in every way; and now, as I look back on the days of my former life of sin, and remember how blindly I would start off on sprees, finding myself sometimes in one place, sometimes in another on coming to myself, and when I reflect on the infinite goodness of God which kept me unharmed through such experiences, I feel that I can never serve Him enough. How I love this Home! It is my earthly home, and sickness alone can ever keep me away from these Saturday night meetings. Each day I ask God for help and strength, and my prayers are never left unanswered."

Mr. O. (ten years steadfast), the oldest graduate of the Home, said : " I too desire to testify to-night to the saving goodness and mercy of Almighty God. Nine years ago, when I first entered this Home, I had lost friends, social standing, and all that makes life worth living. I was separated even from my own family. Walking out one day, while an inmate of the Home, I met my wife, who had only words of reproach for me. She would not, she *could not*, believe that I had really reformed, and it was a long while before she could trust the reality of my conversion. To-day, however, God has given back to me all that I had lost, and restored to me the love and respect of all those from whom I had become estranged. I have not tasted liquor since I left here, and I feel that through Divine goodness I am saved to the uttermost."

Mr. E. (over three years saved by Jesus) : " Let me add my testimony to the love and mercy of God and of His Son Jesus Christ. When I was brought here, a wreck in every respect, it was, as my friends firmly believed, only to die. As to my reforming from my evil habits, such an event was not even taken into consideration. I was adjudged a hopeless drunkard by all, even by my own brother. Thanks to God and this blessed Home as an instrument, I have succeeded in gaining back the good opinions I had lost. When at the point of death, and wandering deliriously in my mind, I heard a voice saying, ' Lo ! I am with you always.' Then all fear of death left me, and I became certain then and there that all was well with me, even should I never rise again from my sick-bed. O brethren ! if you trust in Christ and accept Him, you need fear no evil ; you can face all dangers. I put my whole trust in Jesus, and ask that you remember me in your prayers."

Mr. S. : " Before I entered this Home I was an outcast and a wanderer, without home or friends. Being obliged in my business to handle liquor constantly, I yielded unresistingly to its influence, until it brought me right down to walking the streets for want of shelter. I had tried hard to become a Christian, but did not know the right way. At last I gave up my situation for the purpose of coming to the Christian Home, and have already cause to bless God I did so. When I leave I shall never seek a situation where rum is handled."

Mr. P. : " Two months ago I entered the Home, a wretched, broken-down object. Soon, however, the Lord took hold of me and made me a new man. I had tried, I suppose, a thousand times, to give up drinking, but I always did so relying on my own strength. Here, however, I learned to look to God for succor, and He has taken away from me all appetite for the liquor which I used to drink from actual love of it. My companions at the shop where I am employed were disposed at first to sneer at me for my change of habits, but knowing the details of my past life, they have ceased to joke about it, and respect me for the principles I now advocate. I put my entire trust in God, and pray that with His help I may stand firm."

Mr. S. : "It is eight months now since I left the Christian Home. When I entered it I was a complete physical wreck, without a gleam of hope or a spark of energy. I was treated so well here that at first I could do nothing but cry whenever I found myself alone. Mr. Bunting showed me the way to salvation, and made the path smooth for me. After leaving the Home I was for five days without a place to sleep, often eating nothing for a day at a time. Then it was that temptation assailed me, and that the sneers of my companions did all it was possible to do to discourage me and cause me to doubt God's mercy. I stood firm, however, and it was not long before I was taken back by my former employer. How can I ever be thankful enough! Those who knew me as I was formerly have changed their opinion of me, and I do my best to deserve their respect. Money could not pay what I owe this Home, and I glorify God for His loving kindness."

Mr. S. : "I testify to-night to my love for Jesus Christ and my admiration for this Home. A little over a year since, drink had brought me down to the lowest depths of degradation, and unfitted me spiritually, morally, and in every other way for this world or the next. Many and many a time in former years I had tried to raise myself, but could not, because I trusted to my personal efforts instead of looking to God for strength. My good resolutions would last perhaps a month, and then social life and good fellowship got the upper hand. At last I was moved to ask God for help, and made up my mind to cast my burdens on the Saviour. He directed me to the Home, and it will be a year on Tuesday since I entered it. I came sick and exhausted, and on my arrival fell into a fever which lasted three or four days. At the end of that time Mr. Bunting sent for me and spoke such words of comfort as I had not heard since my mother talked to me when I was a child. They touched my heart ; they opened up new prospects of peace and happiness, and through them I was led to Jesus. Ever since that time I have remained steadfast, and put my whole trust in God. During my business hours, and in the midst even of my business cares, I find always time to utter a few words of prayer for protection and guidance. Notwithstanding the sneers and gibes with which I am frequently saluted when down town, because of my conversion, I persevere in my efforts to do right, and having the elements of strength from a higher Power, I have, thank God, put the enemy under my feet. As a graduate of this Home, I feel an interest in every one of its members, and have helped to direct one or more of my friends in the right way by leading them here, knowing the benefits I have derived from the same source. No earthly remedy can save us from the curse of intemperance. Our only hope is in Christ, and to Him only can we look for salvation."

Mr. S.: " I thank God that the religion of Christ costs nothing. If it had seven years ago, I would not be here to-night, cleansed by the blood of Jesus. When I accepted Him I settled the question of salvation once for all, and am ready to serve Him for the rest of my days. remembering that "they who trust in the Lord shall be as Mount Zion." The hope of future happiness in eternity upholds me in the midst of this world's troubles and trials. Some day, brethren, if we live up to the professions we have made to-night, we shall meet in heaven and talk over all these matters. In the meantime, I am content to believe that I am saved through faith in Christ."

Our religious services consist of morning Bible readings, afternoon prayer meetings, expressly for the members of the Home ; a Tuesday and Saturday evening meeting, with testimony and prayer. The Saturday evening meeting is public, to which all are invited. There is also a Thursday evening meeting conducted by one of the members, and I feel that Mr. Stainback, under whose special charge the meeting has been held for the last four years, deserves our confidence and esteem for his unwavering earnestness in conducting this service. Souls have been saved, and often I find that the first impressions on a member of the Home were made in this meeting.

WORKERS WANTED.

We do seek assistance in this great work of regeneration from outside our Home, and will you not, my dear brother in Christ, preach as you stand in your pulpit the doctrine of St. Paul, "It is good neither to eat flesh, nor to drink wine, nor anything whereby thy brother stumbleth or is offended, or is made weak." We then who are strong ought to bear the infirmities of the weak, and not to please ourselves. And also, my sister or brother in Christ, will you not, by your example, try to lead the poor tempted one from this sin by proclaiming yourself on the side of total abstinence. If Christians would do this, how soon would drunkenness in our city be unknown, and this great evil and sin be classed as belonging to the things of the dark ages. Let Christian life and example put a blasting curse on this most terrible of sins.

Every possible influence should be brought to bear against intemperance and its causes; and foremost in the van of total abstinence workers should be the Christian mothers, wives and sisters of America. It is a striking but too often forgotten fact that the power for good which lies within the reach of young women especially, has never been exercised to a degree in any way as great as the power for evil which follows those of their number who are thoughtless, frivolous or worse. Writing on the subject of the interest which should be taken in temperance reform by young ladies, Miss Carrie Scofield, of Wheeling, thus expressed herself a short time ago :

"The question that puzzles me most in this work is how to gain the interest of our society girls, the girls who in many respects seem to have the most influence, the girls of education, polish and refinement—not but what we have a great many just such girls, but as a rule—understand, now, I am speaking of them as a class—they are the ones to hold themselves aloof. It seems to me this is the way it is in Wheeling, the young ladies who have the most influence seeming to have the least interest, but with a little careful and delicate management I am sure we will soon have a union started there. The time is coming and it is not very far off, either, when we shall be mighty proud to be numbered among the temperance girls of the nineteenth century. But do we as yet begin to realize what a power is ours? Has the startling truth yet dawned upon us that we make society just what it is? Have we yet thought what a mighty revolution there would be if we American girls would stand as one grand army and demand of the young men with whom we associate the same sweetness and purity of life that they demand of us? Oh, that the full meaning of those old, beautiful words of Charles Kingsley could be indelibly written upon the heart of every living girl : 'Be good, sweet maid, and let who will be clever; do noble things, not dream them all day long, and so make life, death and that vast forever, one grand, sweet song.' Girls, can there be anything grander than a true woman? Is not this the birthright of each and every one of us? Do we begin to realize this? Do we ever think of living up to our high privilege? Are the most important things in life to us the elegance with which we dress, the ease and grace with which we dance, the sprightliness and volubility with which we pour forth small talk, the quickness and deftness with which we play at cards, the number of beaux we have? Or are they the fullness of love, joy, peace, long suffering, gentleness, goodness, faith, meekness, temperance? Girls, whoever we are, wherever we are, whatever we are, God demands us to be true."

Yes, God demands you, women of our land to be true to Him, to all that is best within you, to all the possibilities for good which are within your grasp; and if the poet's dream be true, that she who rocks the cradle rules the world, then tremendous is the responsibility which rests upon you if you fail in asserting your right to be wooed and wed only by men of Christian temperance and godlike cleanliness.

A TRIBUTE TO MRS. BUNTING.

Perhaps no woman lives to-day with a more practical knowledge of the power of her sex to win souls to Christ from the ranks of intemperate men than the one to whom it is fitting in this place I should pay the tribute of a heart, soothed, consoled, cheered and blessed beyond its deservings, by her gentle, loving, Christian ministrations. Beside me in every good work, my wife knows what it is bring all the resources of feminine influence to bear against the demon of strong drink. She knows what it is to have lived with a husband when his heart was closed to Jesus, and she realizes all the joy, the peace, the comfort that poured in when the soul of the prodigal relinquished the husks of sin for the manna of a salvation, bountiful, beautiful, plentiful as God. When upon her anxious ear fell the sweet tidings that I had given myself to Christ, she rejoiced in the Lord, and since I undertook to cultivate this special field of duty, she has been with me, bearing many of my crosses, weaving many of my crowns, lightening my burdens, and proving herself in truth and in deed a helpmate in grace as in nature. It has been her privilege to lead the meetings in hymns of praise day after day. Upon her delicate but ever willing shoulders rested many a responsibility during months of weary sickness to me. At all times she finds it a source of gladsome growth in the Christian life to minister words of hope to souls darkened by the consequences of sin, to pour the balm of Christian charity upon hearts saddened by the results of disobedience to God's law; and with all her knowledge of the pain, the woe, the untold misery caused by drunkenness, she joins her voice with mine in supplicating Christian women to enter upon a crusade against rum, and never to relax their endeavors until the curse of alcoholism is banished and buried forever in the depths of the bottomless pit. She knows that the foul appetite for strong drink in man has ruined the lives of more women, blasted more homes for them, weighted them with more sorrows, blighted for them more fortunes and cast upon them more shame, more pinching hardship and brutality than any other evil on the face of this sin-cursed earth. She knows that there are thousands of women who are widows to-day, borne down by almost unbearable loads of hopeless needs, because their husbands have been slain by alcohol, or morphine, cocaine or opium. She knows that throughout the land there are thousands of homes in which wives endure lives of unspeakable torture because the rum-fiend has seized those who swore to love them. She knows all this and more than pen or word of mine can describe. But she also knows that there is a Jesus who is mighty to save. She knows that when all heaven and earth seem to have forever abandoned the intemperate slave of Satan, he has but to look to Jesus, and that salvation is as sure as God, and everlasting as His throne. The more than two thousand men who have been in this Home, will bear me out in saying that her presence conduces to their improvement. They feel her sympathetic spirit, and as I, with them, acknowledge a sense of indebtedness to her for the assistance given in my round of official duties, I bless God that He gave her to me, to tread with me His loving ways on earth and hereafter to share the supernal joys He has prepared for those who serve Him to the end.

EVILS OF MODERATE DRINKING.

Can any one examine the Court records for even a month, and see the instances of crime, suffering and want caused by drunkenness, and feel for one moment that in God's sight all is being done that can be done to drive the curse from our midst? Who will advocate even a moderate use of this deadliest of all poisons to the human family at large? Moderate drinking, I know, is a subject which is very distasteful to that class of society, who, among themselves, are above suspicion in this practice. It is when we condemn the practice, that we are called unwise and fanatical, by those, (who in their own opinion) never indulge to excess. If we were to contend for the suppression of the sale of intoxicants, many professed Christians would unite with us, but when we deal a blow against the sin of moderate drinking, we offend many who occupy high positions in Church and State.

Moderate drinking in the home circle exposes the weak to a danger which even the strong are rarely able to resist. Moderate drinking is a domestic indulgence and that which gives pleasure to the parents cannot be refused to the children. Therefore around the family table the first sip of wine is taken. Oh, the soul and body destroying influence thus inflicted upon the unsuspecting child!

Yes, in Christian homes in this city the children are allowed to tipple, and in this, parents are allowing that work to be done at which the Evil One smiles. What the child sees the father or mother do, it of course will do with unsuspecting confidence. Oh, father, mother, do you know to what this practice that you are guilty of, will lead your unsuspecting boy? During the period in which the child's character is forming, he knows of no kingdom but his father's home. Is not the father's word taken on every question? Will not your example influence that boy in his future intercourse with the world? Now it is this far reaching influence connected with domestic drinking customs which sends forth from so many families a drunkard.

Trace, if you will, this evil of intemperance, and it will be found that its origin was the social custom of moderate drinking in the home circle.

A party assembles; father and mother, sister and brother, friends and little ones are all there. The voice of laughter is heard with the song, the jests, the interchanges of social affection. And now the wine is passed, and the contents of the glass are but tasted, yet is there not danger here? Harmless as it all seems, yet the most deadly work is being done; an influence unseen but subtle, is entwining itself like a serpent about the affections of those young hearts, and slowly but surely it will bring them under a most degrading bondage.

Tell me, fathers and mothers, are you not teaching your children to drink, and that they can do it because it is fashionable and safe, and because you drink moderately?

Is not an appetite being formed in them which may yet bring sorrow and disgrace upon your unsuspecting head? And will you, in future years, be willing to admit that it was through your example that your child became a drunkard?

Many of our applications tell this story. When asked the question, "When and where did you first drink of the intoxicating cup?" the answer has been, "At my father's table."

Oh, you may try and deny the fact, but the boy knows where he took his first drink. If you allow your children to drink, and sanction the custom in your own house, rest assured that between your sanction and the mistaken kindness of friends, an appetite will be created which all the remonstrances of friends and respect for character and standing may not be able to control.

Let me give you a true story. The son of a highly respectable family was found to be dishonest and was sent to prison. His father visited the prison. Dissipation had done its work, yet the father knew his boy. "And what do you think of yourself now, father?" said the son, as he stood in the far corner of his cell. "Think of myself, my son," said the father; " what do you mean?" "The glass of wine you first gave me is the cause of it all. But for your example I never should have drank. The wine at the table at home first gave me the impression that there could be no harm in my indulging at a friend's table also."

Such are the fruits of moderate drinking as daily practiced in this city at the family table. Shall we be still, or shall we continue exposing the evils of this most pernicious system? Will not the Church join with us in trying to drive this curse from our homes?

THE HIGH LICENSE FALLACY.

In special fitness with the agreement of view that moderate drinking is an evil, springs up an idea that the manufacture and sale of alcoholic stimulants should be forever prohibited. But say some unthinking friends of temperance, " It is useless to try to prohibit the sale of intoxicants wholly. It cannot be done. Is it not much better to have half as many saloons as now exist, and receive a high revenue from them, rather than try to prohibit and fail in trying?"

No! no! Every time we would say no! Because we cannot banish all these rum shops at once, for the reason that there is evil legislation, and that officers of the law fail to do their even now inadequate duty, will we say, "We will sanction crime by having men pay a large amount of money so that they by law shall be permitted to commit crime?"

Well do I remember, in my youthful days, how I argued with my aged step-father, as I saw him year after year go to the ballot-box and deposit his anti-slavery ballot. I tried to convince him that his vote was just thrown away. He, however, still continued, as he said it was a principle that instigated him to do it. As men argue now in relation to Prohibition, so I argued on the anti-slavery question. It was not many years before I saw his premises of right prevail, and the party in power fade away, and at last go out of existence.

So with this great moral question of to-day. The cloud may be no larger than a man's hand as seen in our horizon, yet it is a cloud, and it has an indication.

But, persist still the good people who are led astray by the High License fallacy, "Establish the High License system and there will be fewer places left to tempt the poor unfortunate drunkard. You are fanatical on the subject—a half loaf is better than no bread." We know some of these people and know them to be honest in their delusion. But this honesty of purpose does not change the inherent character of their act and its consequences, and we say to them in reply, Why license at all? If it be right to sell rum then let men sell it as they do any other article of merchandise. Let them sell it as milk is sold, or shoes, or dry-goods, or bread and meat.

NO COMPROMISE WITH SATAN.

Too fanatical; yes, thank God, if to have and to hold no compromise with the devil be fanatical, I am a fanatic. If to have registered a vow before God; with His angels; my own heart; my wife; my friends; hundreds of men, redeemed by Jesus, using me as His agent, and my fellow-citizens generally, as witnesses of that vow, to wage relentless war against the rum-fiend, and with the sword of the Spirit, protected in the armor of Christ, and my Creator, "as my strong rock for a house of defence," to strive against alcohol with every energy of which I am capable, if this be fanatical, then I am—and glory in it—a fanatic.

Fanatical!

Ah! Look upon things as they are and then let me ask how can you "long halt between two opinions," and when will people learn that liquor-selling, it matters not how high the license, cannot be made to pay a hundredth part of its own expenses. As you look upon your son, the pride of your heart, can you place any amount, however high, before the city or town, as a license for a place where this boy by your vote can get his dram legally and become a drunkard !

What would be a fair compensation for the ruin of your boy? Your poor wife dies of a broken heart; your son at last fills a drunkard's grave, and you say it is better to have a half loaf than no bread. What do you mean? Does not your own flesh and blood pay the High License that you vote to be granted? How can honest Christian men be so blinded. In the name of humanity we ask you to stop and think before you further go.

RATHER FREE RUM THAN LICENSE CRIME.

No! by far let it be free rum rather than by vote license crime. This liquor license scheme is not a scheme to banish alcohol, but to legalize and perpetuate it. Are we not guilty, and do we not become parties to this soul and body destroying traffic, and receive a portion of this blood money that is paid into the treasury of State, city, or town, when we vote for license, high or low? It would seem that the whole scheme is a trick of the devil, and yet as an angel of light he appears to many a good honest heart, and would deceive by these devices and schemes.

What would you think if a petition should be handed you to sign, permitting a crime to be enacted in your town for money, the petitioners saying: "It will be done, and we want you to sanction it, and we will pay for the privilege. It will be done if you do not sign our petition, only we place this before you to give you to understand that if we can be permitted to do this by law, we will act as detectives to prevent those who are not legally permitted to do the things which we ask to be allowed the privilege of doing?" Why, there is not a man with the least honor who would listen to such an argument for a moment !

Now my friends let me ask, what is being done to-day in this community in this regard? Are we not sanctioning crime when we vote for those who prolong the reign of King Alcohol by license, high or low? And is not license, high or low, nothing more nor less than granting a man a permit to pay into the treasury so much money, and by thus paying the amount stipulated, secure the legal right to make a father, husband, or son a miserable drunkard—permitting these rumsellers to rob wives and children of the food and raiment necessary to preserve their lives, and bring sickness and death to innocent men?

This is the precise status of the license question and all this to license a traffic frowned upon by God, and held to be a business that no honorable man can engage in! Is it not a fact that the city or town which licenses men to sell rum, commissions them to destroy the character and standing of their customers? If the dealer by a license sells rum to a husband, and he goes to his home maddened by that rum, and there in his frenzy strikes down his helpless wife, we ask if God does not hold that rumseller as being accessory to that murder? If so, does He not hold the voter responsible for giving that rumseller a right to sell by law that which has made the husband a murderer? Should not the law hold that rumseller responsible? If so, should not the same law condemn the man that made it lawful to sell by license? Answer this, you men who advocate high or low license.

God has said, and it will prove itself true, "Though the wicked join hand in hand, they shall not go unpunished."

SUPPRESS IT UTTERLY.

Understanding and realizing the evils which flow from the manufacture and sale of alcoholic stimulants, is it not time for us to ask, Is there one reasoning, intelligent Christian man who will try to uphold this traffic as at all necessary to our prosperity, peace, or happiness? Has not this trade in alcoholic spirits from the first been considered a debasing and dishonorable business? Are not its patrons constantly in the hands of the courts of our land ? Whenever spoken of, has not this business been denounced by the better class as a dangerous thing to be allowed in any community? And from what we have seen, from what we have suffered, from what we have known about it, has not the time arrived when we should as one man declare that forbearance ceases to be a virtue, and as our long-suffering is exhausted, and reformation seems hopeless, is it not time that we demand by law that this curse be driven from our country, that the license law be abolished, that all kinds of limitations and restrictions be ended, and the sale of alcoholic stimulants as a beverage be prohibited by the laws of our land ?

Are we not square in our premises ? Is not the purport of the law which has been established in these United States the protection of our interests and general welfare ? Now, this law is supposed to protect everything that would promote our general welfare, and to suspend everything antagonistic to its best interests. Does it not seek to guard life, liberty, property, and morals from every influence hostile to them? It matters not what may be the cause, so long as it interferes with these, would it not be considered amenable to the law in any of our courts of justice ?

Then do not find fault or censure us for the stand we have taken against this vile and insidious destroyer of all that is good and holy, peaceful and lovely on the face of this broad earth. It is upon this ground we make our declaration. It is upon this broad principle we make our defense, and call upon our legislatures to pass such laws as shall protect us from the evils of this abominable traffic which is the maelstrom that will soon engulf this nation.

RUM THE COMMON ENEMY.

While this traffic is allowed in our midst, life is in danger ; railway trains are driven to destruction, elegant steamers, freighted with passengers, on a bright moonlight night (as on Vineyard Sound, for example) are driven upon the rocks on our coast; and the account of murders committed through rum are perfectly astounding. One hundred thousand miserable wretches, maddened by the curse, are buried from sight yearly; and still in this boasted Christian land we say we cannot help it and therefore make all this legal by a license. Should not such words as these, coming from true American hearts, cause us to hide our faces? Have we not a right to claim protection for our property and that of our friends? What protection has the wife of a drunkard to-day? Her home is turned over to legalized rum-sellers, her children are sent to the asylums of our country, and the bloated wretch of a husband is arrested from time to time, and the city finds him a home in her public institutions, while the poor wife is left to drag out a miserable existence in some basement or attic. And yet we boast of our land of liberty; and with all this these destroyers of all that is good call for their rights to be respected, and ask that laws may be so made that they shall feel secure in their hellish traffic.

Now, if it be right by law to punish crime, is it not right by law to prevent crime? Are we to have a law for the suppression of crime and another law legalizing the originators of crime? Shall we by law imprison the man who, infatuated by this demon, rum, beats his wife and children, and not try rather to seize upon the demon before it is allowed to enter the man?

And now, after all this, we would ask any fair-minded person if we are hot headed, extravagant in our demonstrations, or can we say anything harsh enough against this disturber of family peace and enemy of civilization? To the rescue then, ye men of principle, and do not let it be said longer in this our native land that it cannot be stopped. Call upon God, ye men of God, and pray without ceasing until this enemy of all good shall be driven from our land.

RUM-SELLING A CRIME.

Many of my readers are aware that there are those who parade before the country as TEMPERANCE LECTURERS—men who declaim against the folly of drunkenness, and yet appear anxious to palliate the crime of rum-selling, excusing the rum-seller, in fact, as a mere deluded individual who does not realize all the horrid enormity of the crime he is committing. Out upon such loud-mouthed apology for reason! Out upon such manifest travesty of common sense! It might be possible for such excuse to find justification in those days when no license was granted by the community to the rum-seller, when no man was expected to pay $500 or any other given sum of money to make the business of rum-selling legitimate before the law and consequently, by a legal fiction at least, quasi-respectable in itself. No man engaged in that accursed traffic in this nineteenth century is ignorant of the fact that he is committing an outrage on civilization, and when any pseudo-temperance advocate talks to me of justice and mercy, for his "brother" the rum-seller, I feel as though such person were going out of his way to plead for a perpetrator of crime equivalent to murder, highway robbery, or theft.

Palliate, plead for, excuse, extenuate the rum-seller and his business if you will, but all your flights of rhetoric cannot hide the deep damnation of his example nor remove the demoralizing effects of his business. His presence is a blasting curse. His business is a withering blight. His gold is coined in the mints of broken hearts. His river of life is fed with woman's tears. His sails are filled with the cries of infancy His seas are strewn with the wrecks of desolated homes. Excuse the rum-seller! No. No man knows better than he that his business, like the upas tree, poisons where it flourishes, withers the tender charities of life, and causes the weeping and the sorrow which refuse to be comforted in their bitter lamentation and woe.

Oh, may God in His mercy look upon the rum-sellers and change their hearts. They are a hell deserving class, whose very self-interest it is to turn the city, State, and Nation into a mass of seething drunkenness, to debase the manhood of the country by changing friends of order into promoters of disorder, upholders of Christian civilization into practical apologists for demoniac barbarism, degrade womanhood, destroy childhood, level the church to erect the jail, abolish the altar to establish the scaffold, and in the very excess of its lustful blasphemy attempt to dethrone God at the call of Satan.

"Ah," cries one, "this is an overdrawn picture." Not so, my friend, for were you to have occupied my position for the past ten years you would have found men—the finished product of the rum-shop—men who had had bright careers and golden prospects before them—so degraded by drunkenness that they would not accept salvation if offered them, would not accept the offer of release though it were accompanied by indubitable proof that they had but to will the acquiescence of their desires with the will of the Master in order to destroy their foul appetite forever. I know that there are drunkards who have reached such a stage of degradation that if they could, and the offer made and proven to them beyond a doubt that the appetite could be overcome or eradicated, would voluntarily continue in their degraded, wretched lives. It is true that such utterly lost and completely abandoned men have not crossed my pathway often. But I have seen enough of them, and of men who had gone perilously close to their lines, to feel that the rum-seller is of all men the one most directly responsible for the largest amount of misery in time and damnation in eternity.

Rum-selling is a crime. Then, in God's name, let us expose and root out the criminals.

IT IS OUR NATIONAL SIN.

It may be suggested that in thus assailing the rum traffic in its every phase I am travelling somewhat out of the record or beyond the scope of this little work. But in presenting facts for Christian thinkers to ponder over, situated as I am at the head of this Home, where the consequences of rum in all their horrible forms are perpetually reminding me of the cause of such ruin, I must cry out—yes, though every other man were dumb I must speak against this agent of Satan in bringing souls to hell. Before me every day are bright intellects, loving hearts, "fine fellows" in every way, all brought low by this accursed thing, and though every pulpit were voiceless, still would I exclaim, O God, do Thou touch my lips with a burning coal, so that, like Thy prophet of old, I may speak Thy message with mightier power day by day. In all the crime that flows from alcoholism the city, State, and Nation share, because in their corporate capacity they handle, they deal with, the provocative cause. No country can boast of greater advantages, no country can offer the rising generation greater facilities for every kind of business pursuit, no country places more abundant resources within the reach of man. No land has been more signally blessed and prospered, and yet it is evident that we are fast travelling the downward road which will soon lead us into a terrible pit. Through false teachings this dreadful evil of intemperance is making frightful inroads upon the morals of our unsuspecting youth. Could we but summon the young men of the country who have entered upon this path of death, and ask of each one the cause of his sad and deplorable condition—-bright prospects clouded, suns seemingly going down forever, no home, no friends, parents broken hearted, if not filling untimely graves, wives and children left to drag out a sad existence—would not each one answer, It is on account of your own personal, civil and national selfishness ; it is because in giving a color of respectability to habits of moderate drinking and to the traffic in strong drink, you have led, thoughtlessly perhaps, but nevertheless assuredly, your unsuspecting children and companions to destruction?

We witness daily, in the case of wretched men who come under our immediate care, the blighting effects of this poison upon the intellect, and this more within the two years last past than ever before. But this is not the saddest part of it. It is the ruin of the soul which is wrought. Oh! the horror that awaits you by whose example or words this curse has fallen upon a brother. "Woe unto them who are mighty to drink wine, and men of strength to consume strong drink." Oh! that those blinded, misled leaders of moderation (?) would pause and reflect before they lead the innocent ones further down to death and hell. Listen while on bended knee you pray, "Lead us not into temptation." Oh! beware, my brother! Are you not, by your own example, leading the thoughtless ones down the slippery, deceitful steps which comparatively few remount, and if they do, in a maimed and crippled condition? Do you boast of your own strength above that of the weaker brother? Because you may be able to drink your brother drunk, will you claim yourself as temperate? God does not judge us by the quantity of liquor we can hold and carry. If it is a sin, you will not go unpunished. Beware, oh my deluded brother, beware! God will hold you to a strict account for your wretched example. The habitual daily use of intoxicants brings on disease and shortens life, and it may therefore be conceded that any use of liquor as a beverage is intemperance, although the effects may not be immediately made manifest in the appearance or conversation of the habitual user.

THE DRUNKARD'S TERRIBLE APPETITE.

It is estimated that in our land we have no less than one million habitual drunkards. Each one of these calls for sympathy and aid, and through the blessing of God they may be saved and gain eternal life. For the sin of drunkenness (and that is just what we must call it) there can be no apology, but that the drunkard's condition is pitiable no one will deny. It matters not what station or standing he may have had, nor how respectable he may have been, his downfall at first may have been gradual, a temperate drinker he may have been called, but the appetite soon is formed. Oh! the drunkard's appetite! It is indeed described as the gnawing of "the worm that never dieth." In this terrible state he calls for help, and at this very hour in his life he would give all for this saving power as he cries, "Oh! wretched man that I am!" Many a man whom we have taken into this Home has sought help in every known way but the sure and only way. They do struggle to overcome appetite. By voluntary commitment they cause themselves to be imprisoned. Sea voyages are taken. They wrestle in agony, and all to no purpose, until God's grace is implanted in their hearts. Will you not come to our help? will you not aid us with your prayers?

The drunkard's appetite is a thing of which no one can have knowledge save the drunkard himself, and even he can only know it by experience rather than by definition. Physicians and scientists have failed in all their efforts to discover precisely of what it consists or the part of the body in which it is located. They have vainly tried to find a suitable name for it, and some have tied it to the palate, others have put it in the stomach, some have placed it in the brain, and others have distributed it among the nerves. But oh! the man who has been once afflicted with this hell-created appetite knows it combines all these together, that it is the whole system of the slave stirred to a mighty frenzy, the entire alcoholized being crying aloud for strong drink. Withhold the drink from him and the intense, burning pangs seem akin to the torments of Dives, who prayed that Lazarus might be sent to dip the tip of his finger in water and cool his burning tongue. The power and bondage of this appetite beggar description. Held fast in its terrible chain, the victims of alcohol are driven to every extremity. Which one of those who have ever mingled freely in barroom society has not seen the bloated sot, shaking from head to foot, taking his very life in his hand, saying to his comrades, "Here goes for another drink." Yes, though he die in the attempt, the wretched man is determined to have it. He jeopardizes soul as well as body rather than endure the craving of the horrid thirst which is consuming him. There are many incidents to prove the terrible nature of this appetite, and I have met with men who declared that though hell itself were to engulf them, yet they should have rum. A gentleman once said to a hopeless drunkard, whom he had in earlier years known as a bright and promising young man: "Why don't you quit this life? Don't you see that you are ruining your family; that you have lost your social position, your property, your health, and that you are going down to a drunkard's grave?" This is the answer that he got: "Do you think you have told me anything new? Do you suppose there is anything in domestic happiness or social position or health even that I have not thought of a hundred times to your once, or that I do not know the value of as well as you do? I have had these things held up to me and held them up myself hundreds of times. I have got them by heart. But they have ceased to mean anything to me. I know what domestic comfort is, but I don't want it. I don't want social position. I don't want the respect of my fellowmen. I don't want money. I don't want health. I want rum. It is the only thing I do want; and when you offer me all these other things you do not tempt me a bit." "What can successfully fight this raging appetite which masters so many men everywhere?" asks the Atlanta (Ga.) "Constitution," and that very ably conducted newspaper replies: "Clearly nothing but the impossibility of getting the stuff for which the drunkard thirsts." But while acknowledging all the advantages possible to Prohibition, our answer to such question is that the grace of God is the only weapon strong enough to strike down the infernal spirit of alcohol. Could any mere human device remove the appetite from the unfortunate man

whose tale of temporal and eternal ruin is told in the following incident :
"One wintry afternoon a trembling man entered a tavern in New Hampshire carrying a small bundle of clothing. Going to the bar, he said :

" Landlord, I'm burning ; give me a glass of gin."

The landlord pointed to a lot of chalk marks, and said: "John, you see that old score; not another drop till that is paid."

The poor wretch glared fiercely at the man behind the bar.

"Landlord, you don't mean that. You have got my farm and horses, you have got my tools. All I have in the world is this little bundle of clothes. Please, landlord, give me for them just one glass of gin."

"I don't want your old clothes," calmly answered the man. "Pay the old score first."

The drunkard staggered back. A gentleman then said :

"What will you give me for enough to buy two glasses of gin ? I see you have a good pair of boots on your feet ; will you give me your boots for twenty cents ?"

The miserable wretch hesitated for one moment, then said :

"Stranger, if I give you the boots, I must go out into the snow barefooted. If I give you the boots, I must freeze to death ; if I don't give them to you, I shall burn to death. Stranger, it is harder to burn to death than to freeze to death. Give me the gin ; you may have the boots."

He sat down and began to draw them off. The gentleman did not, however, intend to take them, but he was testing the terrible appetite. Others were looking on, and they said the man should have his gin. They supplied him liberally, and he drank all he could and took the rest away. When night came he had drunk the last drop, and he went to sleep in a barn. The frost king came and took the man in his arms. The next morning his body was found dead in the barn ; and, O God ! I shudder as I ask myself, Where was his immortal soul?

To men it may seem impossible to convert such slaves of appetite, but with God not so, for to Him all things are possible. While these men are suffered to live as they do, they are only a curse to the community. The moment they are regenerated society is relieved of its greatest burden. "And what can I do?" you may ask. By your example, my Christian brother, you may prove "a savour of life unto life, or of death unto death." Cease, then, setting an example which may prove their ruin. Of course some are deceived, and in deceiving themselves have misled those who have labored with and for them. This must be so to some extent; but the number is small, as our statistics will show, of those who, having remained for some time in the Christian Home, go back to their evil ways.

In this place they have learned that there is a way to the goal of emancipation. Here they have realized that—despite so-called scientists who attempt to show that because their finite capacity cannot discover a medicine to cure this appetite, there is therefore a limit to the power of Jesus, the Infinite—still there is healing for their bruised beings in the only remedy for sin, the Gospel of the Son of God.

Yes, praise His dear name forever, "He breaks the power of cancelled sin and sets the prisoner free." In this Home, after years of simple but implicit trust as their only safeguard, men rise from their seats week after week—the scholar from his books, the laborer from his load—this one from a group of anxious clients—that one from out a throng of confiding patients—this one from the banking house—that one from the counter—this one from his ledger—that one from the factory—this one from the pulpit—that one from the pew—all proclaiming their freedom, rejoicing in their liberation, purged of every stain, clean as new-born babes from every taint of rum and tobacco, opium or morphine.

Yes, glory to Jesus! He saves to the uttermost, and enables each dear, precious, redeemed soul to cry with Saint Paul, "I am not ashamed of the Gospel of Christ; for it is the power of God unto salvation to every one that believeth."

THE MORPHINE OR OPIUM HABIT.

[These pages, dealing with the morphine habit, have been contributed by a physician who was rescued in the Christian Home.—ED.]

This pernicious habit, destructive alike of all physical energy and of all moral character, is fearfully on the increase. Professor Ball, M.D., of the Paris Faculty of Medicine, estimates that there are to-day hundreds of thousands of human beings enslaved by it.

It is here proposed to notice briefly; (*a.*) The conditions leading to the use of morphine ; (*b.*) The effects of its abuse ; (*c.*) The effects of abstinence ; and, finally, the method and process of cure.

It is necessary to premise that the inducements to the use of opiates, and the facilities for it, have very much increased with the substitution of morphine for opium, and of the hypodermic method for ingestion, say a period of twenty-five years ago.*

(*a.*) The *conditions* leading to the habitual use of morphine are generally either physical pain or mental suffering, conjoined with *a given temperament* which may be called "constitutional exhaustion"—the *neurasthenia* of pathologists. Where this temperament is lacking morphine gives no satisfaction, but the contrary. Hence in such cases the habit is never formed. The drug may be tolerated by them as a remedy for acute transient suffering; but beyond that it is not borne.

It is far otherwise with the neurasthenic subject. These subjects are very generally the offspring of intemperate parents; one or both parents having been alcoholic subjects, and while utterly denying the assertion that drunkenness is hereditary, yet it is a fact that the nervous system of such offspring is generally hypersensitive, and the individual, often talented, lacks all *sustained power*. Eventually by some accident the sustaining effects of morphine are experienced, and from that moment the snare is sprung; the individual is a predestined *habitué*.

The authority already cited well remarks that the morphine habit, in its wider sense, "was literally created by a physician;" for although there were victims of opium and morphine before Prof. Wood's invention of the hypodermic method; yet the introduction of that method marks an era in the history of this vice. From that date the habitual use of the drug has increased many hundred fold, and has attained a fearful rate of progress by which it threatens to sweep away multitudes of bright intellects, and to desolate many homes.

(*b.*) For although the first effects of this habit are exhilarating, and many subjects experience no ill effects for a lapse of weeks or even months, yet the inexorable demand of the system for increasing doses, in

---

* The introduction of morphine as a drug was a good deal earlier; but, owing to its expensiveness, it was not in general use until the invention of the hypodermic method.

order to maintain the desired relief from suffering, sooner or later results in paralysis of the will-power, perversion of the moral feelings, insensibility to natural ties and obligations, intellectual dulness, lethargy, dyspepsia, neuralgia, hallucinations,—and finally Bright's disease of the kidneys, and death—moreover, if any acute disease supervene before the kidneys are disorganized, the recuperative powers of nature are so undermined that the chances of recovery are greatly diminished.

The perversion of moral feeling and consequent change of character is in nothing more evident than in the singular mendacity of these unhappy cases. The same individuals who were previously truthful and honorable will now lie without compunction, and in other ways exhibit the loss of all sense of honor.

This deplorable picture is not overdrawn. *Morphinomania* is the term by which medical men have named this unhappy condition. It becomes indeed a mania, a derangement of mind and of sensibility, as well as of body. But we ask in all charity, Is it not the victim's own fault? Some indeed fall into the snare before they know its danger or to what ruin it surely leads. But all such who are of ordinary intelligence soon know that they are committing sin against their own bodies and their own souls, as well as against friendships and society and God. And they know that if they cannot deliver themselves it is their duty with all speed to seek deliverance by means of such helpers as may be found.

This brings us to our third point, (*c.*) *The effects of abstinence.* An individual once under the power of considerable doses, even for a few weeks only, has not the power of his own will to deliver himself. If he undertakes to diminish the daily allowance, and is still possessed of some determination, he may sustain the consequent suffering for a few days— one, two, or three—but he is sure soon to lose his self-control; as surely as a patient undergoing a surgical operation without anæsthetics. He may endure the torture bravely for a limited time, but the limit is soon passed: judgment and self-control are alike dethroned.

If now, by submitting to the stronger control of others, the diminution of doses is persevered in, the chief suffering, aside from loss of appetite and inability to sleep, is quite generally from an indefinable sense of anguish referred to the epigastrium, but probably seated in the solar plexus of nerves. This sense of anguish is accompanied by a painful exhaustion pervading the whole system, and all this, together with the protracted insomnia, renders the condition deplorable enough.

That there are permanent recoveries by this method, if proper adjuvants are used, is not denied ; but we are inclined to think they are few. The system becomes, so to speak, exasperated by the slow torture of the gradual withdrawal of an accustomed stimulant ; judgment and self-control are reduced to the last degree, and during the long period of convalescence yet to follow the determination yields.

This brings us to our final point—*the method and process of cure*.

In the Christian Home the first step taken is the abrupt and absolute withdrawal of morphine and every other opiate from these cases when undertaking a cure. As in surgical practice, so here : the quick, sharp method is the way of mercy as well as of science. The immediate shock is greater, but the sum of suffering is far less, and the desired end is far sooner attained and more surely.

This, however, presupposes that the case is placed under favorable conditions. It is not possible to pursue this plan successfully in one's own home and among friends. Isolated in some kindly institution, having suitable apartments, with nursing and medical attendance, the recovery, after a conflict of a few days, is often rapid. Absolute seclusion, with rest in bed, is essential for a period of ten or twelve days. During the first five or six days of this period the suffering is greatly diminished by judicious medical appliance. There are two objects in view, namely, to tone and to quiet the shaken and suffering nerves. It would be not only cruel, but also unsafe, to leave the sufferer without such relief as medical art and nursing can minister.

At the end of five or six days medical means may generally be diminished or withdrawn ; but the patient should still rest absolutely in bed, and under the eye of medical attendant and nurses.

At or about the tenth or twelfth day he may leave his bed, and by the end of the third or fourth week he will realize that he is fully convalescent; but his absolute quarantine should be prolonged considerably beyond this period, until self-control is fully established.

The powers of nutrition are so much impaired that the quality as well as the quantity of food taken requires attention. Milk in considerable quantity, and frequently, is an important addition to the diet.

In the Christian Home this is the method pursued, together with a kind and persistent moral influence, with faithful testimony that no man once enslaved by a destroying appetite is ever safe except he has the love of God in his heart and the filial fear of God before his eyes. A life of trust and of prayer is the only sufficient refuge from the pursuing evil. The Lord Jesus is the Saviour of the body as well as of the soul.

HOW COCAINE ENSLAVES.

To a non-professional man such as I am it cannot be otherwise than astounding to learn how many of the very men who ought to best know the results of using drugs, whether as stimulant, narcotic, or anæsthetic, resort to such means of debasing their bodies and destroying their souls, and yet it is precisely from the ranks of the medical profession, or from among those who are related to it, such as pharmaceutists and apothecaries, come the larger number of persons who in this Home have sought release from their horrid appetite. As in cases of opium or morphine-using so in cocaine, whatever observation we have had has been furnished chiefly from the medical profession. This cocaine habit seems to be the very acme of pernicious appetite, and reduces its slave to the lowest depths, changing the stalwart form, the ruddy cheek, the erect bearing, into an emaciated, hollow-eyed, bilious-faced, flat-chested, helpless limp of humanity—a very caricature of manhood, with a look like a hunted beast, the shrunken frame trembling, the will-power utterly wrecked, every lingering sense of personal honor and cleanliness destroyed, and but one madding desire—to use the awful drug at all cost, at any peril.

It is difficult for one who is not a physician and has never had the misfortune of being under the power of this peculiar appetite, to properly describe the results of using cocaine, and many of their chief characteristics, as narrated by its victims, resemble those which follow the use of opium or morphine. From a state of melancholy and wretchedness the user of cocaine, under the influences of constantly increasing doses, arrives at a stage of dreamy intoxication, half delirium—a languorous pleasure, like listening to enchanting music. Then other sensations break in, and there are reveries of sadness, feverish, unsatisfied longings, weird terrors of nameless things, vague apprehensions of unreal dangers. Gradually the victim's hell is reached as the effects of the dose begin to wear away. From a semi-delirous delight—from a nameless sadness the wretch falls into a horrible depression, an atmosphere of darkness and suffocation, the air peopled with threatening shapes until the unfortunate one wants to die. So much as is known professionally of cocaine it is not narcotic but a stimulant, with subsequent depression, and is a local anæsthetic of power and beneficent result in the hands of a skilful and prudent surgeon. All the slavery in which the laudanum, opium or morphine user is held (and the cocaine victim invariably uses morphine alternately) keeps fast hold on the bondman of cocaine and to help bring about release from its deadening grasp our methods in no wise differ from those we apply to the victims of other terrible drugs. Seclusion, absolute withdrawal of the drug, nutritious food, care and rest for the body, reliance on Jesus, the Great Physician, to heal the soul, placing the strength of His mighty will against the trembling will of the victim and sustaining the weak one by "the living bread which came down from heaven"—these are our means and with them we are ever ready to battle against the demon, hurl him down and cry victory in the name of the Lord.

THE CHLORAL HABIT: ITS EFFECTS.

After an apparent forgetfulness of the once famous hydrate of chloral, recent fatal doses seem to have awakened public interest in this powerful drug. I need not, says a well informed writer, describe it further than to say it is a salt of burning, pungent taste, having a great affinity for water; it is closely allied to chloroform, into which it is supposed to be changed in the blood. In small doses it is stimulant and anti-spasmodic, in larger narcotic, while an overdose quickly causes death by paralyzing the respiratory nerves. Like opium, the dose must be constantly increased to keep up the same effects. The stimulation, however, is not like that caused by opium or alcohol; it is not exhilarating, and does not incite to action, either mentally or bodily. But the subject of the influence rises for a time above all his cares or sorrows or fatigue, and seems to look on life through the medium of a rose-tinted glass. But while care and sorrow are forgotten, and a dreamy sense of perfect ease, comfort and happiness take their place, all affection and love are likewise banished. The subject is apathetic, and cares for nothing, save his own sense of comfort. In this state the confirmed chloral-taker would stand by the deathbed of his nearest and dearest a passive spectator. If the same dose is repeated the victim either sleeps or shows signs of intoxication.

"I know from experience," said a gentleman to a friend, "the work that chloral does. For many months on my return from business I found my poor wife drunk, and my children, who used to be so merry, silent and unhappy. But there was no smell of intoxicating liquor in the room or even about her breath, and all my efforts to unravel the mystery were unavailing. But one evening after tea she dropped from her chair while trying to speak to me—dropped like a log on the floor, and I carried her to bed. Her face was red and swollen, her lips blue; her arms and legs were marble cold, even hard; she had no pulse at the wrist, but breathed as quietly as an infant. I sat beside her all that long night. Toward morning the sleep was broken by moans and deep catching sighs, and when she awoke it was terrible to look upon her sufferings and agony. From the doctor's lips I first heard the name of chloral. She is now a nervous imbecile, and must, I fear, soon succumb to her infirmities." Yes, chloral is set moving in society, and thousands annually fall beneath its wheels. Let any reader of these words ask any wholesale chemist, and he will be told that large quantities of this dangerous drug are annually imported (from Germany and other parts) which are not prescribed by medical men, but taken as stimulants by the people themselves. Hydrate of chloral in every shape—unless exhibited by the hands of a skilled practitioner—is an insidious and fatal poison. It is more tempting than alcohol, more insidious than opium, and more terrible in its effects than either. An opium-eater, baneful though the practice is, has been known to live to a goodly age; no chloral-taker ever lasted over three years.

For all these pernicious appetites we recommend the same treatment as applied so successfully hitherto, and we promise the slaves of these drugs that if they look to Jesus He will save them, for He says, "He is able also to save them to the uttermost that come unto God by Him." We have had victims of all these appetites here, even within the past year—opium, morphine, chloral and cocaine alike—and were successful in each case, not through our own power, but in and through Him who has taught us to know that he is safe and will be healed forever who cries, "Not my will, but thine, O God, be done."

A STAY HERE NOURISHES THE SOUL.

It may be asked why it is that a stay in this Institution is so conducive to effectual reformation, and, natural as I admit the question to be, just as easy do I find its answer. It is because here a man has an opportunity of entering into himself which it is scarcely possible for him to find while he remain amid the busy scenes of the outside or business world. In a round of idle amusements, or while permitting the mind to be occupied by business cares, it is beyond any man's power to devote the time necessary to the one thing essential—securing his salvation. Secluded from distracting scenes, the single object pursued by an inmate of this Home is how best he may find redemption from his besetting sin, and he is directed to put forth one earnest, undivided effort to use all means afforded to effect permanent reformation and a Christian life. Then it is that Christ enters into the heart which has learned to trust Him fully, and then it is a man discovers that when he makes a complete surrender to his Saviour he is saved to the uttermost. A fountain of living water springs up in the heart of the redeemed sinner, and as he blesses the Lord for the salvation given him he finds that the taste of the waters of life has taken away, and forever, the thirst for strong drink.

God keeps those who put their trust in Him, and those who seek Him earnestly leave this Home fully emancipated from the thrall of drunkenness. His grace is our cure-all—no nostrums, no patented humbugs, but Jesus, and Jesus only, places the man who submits his will and his body to our treatment upon the solid rock of an everlasting salvation.

How many discouragements are to be found by the honest, devout teacher as he endeavors from time to time to lead the lost sinner out of his thraldom of sin. So many hindrances are presented, so many "hard sayings," as they are thought to be—so many difficulties are set up that if the teacher were not filled with Divine grace many times he would be wholly discouraged and give up in despair. Not that I would be found complaining; but I must say that there is not a work in this city that has so many obstacles to contend with as this salvation of the intemperate.

A CONFIDENT FAITH REQUIRED.

We find so many unbelievers that at times our heart fails us. Many men enter our Home sick and tired of their lives of sin, and as the Gospel plan is presented it is eagerly grasped, fully satisfied that at last they have found the real panacea for all their ills. Their better judgment is convinced of the truth as here set forth until a word is spoken in relation to some darling sin; then at once we hear, "This is a hard saying;" at once a murmur is heard. From this time no further advancement is made in the Christian life, and these poor souls on whom all eyes have been turned and hearts have been made to rejoice over, are filled with sadness and discouragement. Then come the chilling words from those whom you had been led to believe were true friends of the cause you were so faithfully laboring in: "I thought it," "I have been thinking it might be just as I see it," "I have feared this falling away," "I have doubted the genuineness of the work," "Many times I have had fearful forebodings"—of what, let me ask? of what? of the truth of God's word? of the unfaithfulness of humanity? Do not, I beg of you, whoever you may be, do not get discouraged. If this were not God's work, and these very truths taught in His word that we are experiencing, you would have reason to fear. But not so. Are we not taught in the 6th chapter of John that there were disciples of our Master who made use of these very words, "This is a hard saying?" Yes, it is plain to any thoughtful reader that we are taught to expect a falling away, a turning back. Is it not taught us plainly by the Master "that it is not every one who saith Lord! Lord!

that shall enter into the kingdom of heaven." But listen; the promise is to the one "that doeth the will of my Father which is in heaven." Is it not plainly taught us here that all disciples are not true believers? Many to-day that are professing to be disciples of Christ, and have their names numbered with the true believers, and are called His disciples, have not, nor did they ever have, the real grace, the saving grace, which is the real gift of God, implanted in their hearts. We need not therefore be disappointed when we find men deceived, when we find men hypocritical. We must expect it; it was so in our blessed Master's time; it will be so in every age. Not all that are swept and garnished have received the saving gifts into their swept and garnished bodies. Many men who come to this Home are convinced of the truth as taught here, and for a time become disciples; but as we have here stated, when the time for the "hard saying" comes, they say, "Who can hear it?" To some, Christ's commandments will always be received as "a hard saying," and thus the evidence is sure they have never had a change of heart, they have never been born of God.

THE SPIRIT QUICKENETH.

It is just one of the many ways in which the natural corruption of man shows itself. "The carnal heart is enmity against God." What else can we expect but this from the unbeliever? Therefore let none be discouraged, although many who profess turn back and go wallowing in the mire of dissipation and sin. If there were not a true and genuine coin you would never see a counterfeit. Let us take God's word, and I know the true believer will not be found saying, "Who can hear it?" "It is the Spirit that quickeneth." It is the Holy Ghost who will give us an understanding heart. By His agency it is first imparted and afterward sustained and kept up, and unless it is imparted we need not expect it to endure. Jesus speaks to a believer, and says, "The words that I speak unto you, they are spirit and they are life." By this He would have us understand that unless His words and teachings are applied to the heart by the Holy Ghost, we need not be disappointed and discouraged when we see men falling who have been members of this Home or of Christian churches. Men may deceive men but

God they cannot deceive. He knows from the beginning who they are that truly believe and are born of God. Therefore be not discouraged oh ye of little faith.

We would earnestly request you to persuade any relative or friend you have who is addicted to intemperance to take a look at this book and thereby ascertain the nature of our work. Tell him that he is invited, earnestly entreated, to forsake his sinful habits, and that we stretch out our hands to assist him, both to leave the ways of sin and misery and to walk in the paths of happiness now and evermore. Urge him in love to avail himself of the opportunity we offer him to regain his manhood. Show him that by thus giving practical evidence of his desire to cast off habits he abhors there is no cause for shame, no place for diffidence, but that all good and true people will sympathize with him in his earnest endeavors to retrieve the past and establish his future on a firmer basis. Endeavor to explain to him that by joining our band of temperance brothers he is not placing himself under any confinement whatever. The man whom we cannot retain in our midst by moral suasion, and by the influences of the surroundings that he will have in the Home, will, we fear, have but little hope for a change of life by any plan now known outside. At the same time, we would beg you to seek that he may realize the fact that it is only through God's grace he can ever hope to effect a radical, a permanent cure. It is to the foot of the Cross we shall direct his footsteps, there to cast off forever the burden of his guilt and misery. Every man who allows liquor on any occasion to get the mastery of him, is termed an inebriate or drunkard. A man who would feel the slightest pang at being compelled to relinquish the use of liquor for a season is in danger of ultimately becoming a drunkard, though he may now be but a moderate drinker; and in order to avoid such a fate it is his sacred duty to become a total abstainer. Many who know they have lost the control of themselves by too frequent indulgence are anxious to conquer the habits they have contracted, have made many resolutions to amend their ways, have been successful in battling the evil in their own strength for a while, but have again yielded to temptation, and at each attempt, losing more of their self-respect, they have drifted farther off on the road of self-indulgence, and sank deeper into the mire. And yet each one exclaims to himself, "What would I not give could I relinquish this terrible habit of intemperance!" Each one sees before his open eyes, with terror, a horrible future, a misspent life, a drunkard's grave.

Our earnest wish is to assist such men to fight the good fight they are anxious to engage in, to lead them on to victory. For this object our Home has been established; here they can live in seclusion for a while, away from all temptation, away from all that could allure them to paths of sin. And after they have regained their self-control, when their minds have become free from the pernicious associations that influence them, we wish to point out to them the way, the only way, for their sure and permanent cure. They know of their own experience that their will power, the combined influence of friends and relatives, were unable to restrain them permanently; perhaps they have given up all hopes of ever being cured. They have forgotten that God so loved the world that He sent His only begotten Son to save us, to save them!—that Christ invites all who are weary and heavy laden to partake of His rest. Rest! have they not craved for rest, longed for it, searched for it, and tried to obtain it by stupefying their brains with alcohol? Rest and peace are offered to them here—the peace of Christ, a peace which passeth all understanding. "Believe on the Lord Jesus Christ and you shall be saved." In this Friend of the friendless, Comforter of the comfortless, Forgiver of the penitent, and Guide of the erring, I find a greatness not to be found in any of the philosophers or teachers of the world. No voice in all the ages thrills me like that which whispers close to my heart, "Come unto me and I will give you rest," to which I answer, This is my Master, and I will follow Him.

A MONUMENT OF MERCY.

Before I close I would say, with the deepest gratitude to God, that I feel He has bestowed a precious gift upon me in the person of my co-worker, Mr. J. L. Pulis. Nine years ago and over I took him from the streets of this city, given up by every one who knew him. He is now my Assistant, and through his teachings scores of lost ones have been saved. Truly is

J. L. PULIS.

he filled with love toward God and his fellow men, always ready and willing to do the Master's bidding. During a protracted illness of months by which I was confined to my bed, God wonderfully blessed his labors. Morning after morning for months he was enabled to sound forth the story of God's love to anxious and attentive listeners.

God has especially blessed him in his work all through the past years, and while he has been enabled to water the souls of the thirsty ones, his own soul has been watered, and a growth in grace has been daily manifested. I know it will be pleasing to all, and read with interest, therefore I give you a part of one of his reports :—

"It is with feelings of great humility and heartfelt gratitude, as the Assistant of Mr. Bunting, in the spiritual work of the Home, that I make an effort to say some few words of encouragement to the friends who have given to this Home their Christian sympathy, love, and support.

"No one on earth has more reason to thank God for this Home, and all connected with it, than myself.

"Here I found a home when every other home was closed against me. Here I found a father and a friend, when father and friends had forsaken me. Here, blessed be God, I found Jesus the mighty to save. Here I found a precious means of grace and holy and hallowed influences, and to this Home, under God, I owe all for what I am as a Christian man. My heart fills with joy when I stand before the world a representative of Jesus and the Christian Home, and for years I have been holding them side by side by my testimony and life.

"As Assistant I have been put in charge of a portion of the devotional exercises of the Home. We have a prayer-meeting every afternoon, all the year round, and let me say, as I lead the meeting, a few things about it. Nearly every member of the Home takes a part in this exercise, and I mention this for it is not a very common thing in prayer-meetings. Every one who has professed to be saved sends out the desires of his heart to God, and while we take to God the very details of our life, both personal and the Home interest, our great desire is that we may be filled with the Holy Spirit ; for nothing is taught more plainly in the Home than that every believing child of God should seek the fulness of the blessing of the Holy Ghost, that we may pray aright, talk aright, and live aright, and in answer to these prayers the entire Home is made the house of God and the very gate of Heaven. Another striking feature of this meeting is the mutual influence. The rich and poor not only meet together, but pray one for the other, obeying the Divine injunction, 'Love as brethren, be courteous, let every one of us please his neighbor for his good to edification.' We are all on one level: 'Let the brother of low degree rejoice in that he is exalted, but the rich in that he is made low.' Here our sympathy, love, and prayers all blend, and make us all one in Christ Jesus, and thus we promote holiness in the hearts of all the members, that they, one by one, may go forth from the Home to live consistent, useful Christian lives, as ornaments to society and pillars in God's church.

"Beside this prayer-meeting, we have one or two other things that give us great encouragement and comfort. When we ask these men to give up their tobacco, you would be astonished to see with what willingness they comply, and what an ardent desire they manifest to rid themselves of this great idol also, as we explain to them that it is a rule of the Home, a safeguard against temptation, and, above all, a very important thing to full consecration, and a step out from darkness into the light and liberty of the Gospel of Jesus Christ; thus leaving them disenthralled from every yoke of bondage.

"Another blessed source of comfort is to find these men always ready to give an answer to every man that asketh a reason for the hope that is within them, and as I have the pleasure of getting a written reason, or testimony, from each one before he leaves the Home, my heart is often made unspeakably happy as I read the depth of love, soundness of mind, and comprehension of the truth in the expression of these newborn babes in Christ, and I am often made to exclaim, Our God is a wonder-working God.

"I could say a great many more grand and good things in connection with the Home and its work, as it comes to my notice in the position I hold; it is enough to say and know that it is God's Home and work: and as such we still hope, trust, and pray that the same interest, love, and prayers will be given it in the future as have been manifested in the past.

"God and His children have dealt very bountifully with us in the days gone by, and goodness and mercy have followed us all these days. Let us send up one universal song of praise by saying, 'Blessing, and honor, and glory, and power, be unto Him who sitteth upon the throne and the Lamb forever and forever. Amen.'"

CAN THE DRUNKARD BE SAVED?

The conversion of Mr. Pulis is perhaps one of the best answers to the above question. When he came here no more miserable being could be found in New York. He was an utter wreck in body and in character, but when he applied for admission he did so with the true spirit. He came with a contrite heart, seeking a Saviour, and He who came to seek and to save lifted the prodigal to His heart, and there he is safe forever. At our first interview Mr. Pulis said: "I have been a constant drinking man for ten years. Five years of that time I was considered a moderate drinker. At the end of my moderate drinking I found that I had become a drunkard. I went quickly from bad to worse. I was soon abandoned by all my friends; the doors of my father's house were closed to me, and every one refused to see me or recognize me as a relative or acquaintance. To-day in a Gospel temperance meeting I heard a testimony from a regenerated man who had been a drunkard longer than myself. He said that there was help for every one. If that is so, I may yet find Christ. I want to be saved, if God will accept such a miserable sinner as I am. I come to this Home hopeless, friendless, and ready to die."

We assured him of the Saviour's promise, "I come to seek and save that which was lost," and soon we had the privilege of hearing his voice in our meetings. I remember after his return, the first day after he was allowed to go out by himself, as he stood up in one of our meetings, his testimony was: "Oh, how I praise God to-night to know that I am free! I passed a liquor saloon to-day, and had no desire to go in. 'I am free! I am free!' I exclaimed, and began singing the Doxology." He is free. Christ has made him free, and he is known all over this city as a monument of God's saving mercy.

My own views on the evils attending the tobacco habit find strong confirmation in the views of Mr. Pulis. He also was under the dominion of the weed as well as of the cup. He praises God for the deliverance granted him, and he treats tobacco and alcohol as twin or almost co-ordinate evils. His matured opinion on the use of tobacco and the results attending such use is best expressed in his own words, and coming from such a source, coming from a man who is, as I well know, taught by the Spirit of God, should be carefully weighed by every one who aspires to Christian practice and desires Christian perfection. He says :

"I think that all well-informed Bible Christians will concede that the arguments against the use of alcohol apply with almost equal force to the use of tobacco, as corrupting, blunting and debasing the moral and spiritual faculties. Those who are in bondage to any evil habit should especially remember the Apostle's declaration to lay aside everything that can hinder or retard our progress, or the progress of others, in the spiritual life. Manifestly, then, any indulgence or practice which is demoralizing in its tendency is a violation of the spirit and letter of God's word, and cannot be tolerated by a consistent Christian.

"The Scriptures teach us that though we may be able to indulge with safety in certain pleasures, yet if our example prove a stumbling-block to a weak brother, we are bound to deny ourselves for his sake, failing which, we deny Christ, our great example. Thus truth, like a two-edged sword, cuts both ways, lopping off the sinful practice or trunk, and the selfish motive which is at the root of the indulgence.

"The sinful waste of money, and the injury to body, mind, and soul involved in even the moderate use of tobacco or alcoholic stimulants, or both, stamps the twin habits with the brand of about equal sinfulness, though the latter is more destructive and far-reaching in its consequences, So bound was I by this tobacco habit, and so enslaved by the appetite, that, even after I had become a Christian, I trembled at the very thought of giving up my idol, notwithstanding the stings of conscience and the strivings of the Holy Spirit. Then came the crisis—Jesus or my idol. The words came to me, 'He that is unjust, let him be unjust still.' 'He that is filthy, let him be filthy still.' I decided for Jesus, and from that moment I became altogether a free man.

"It seems to me that all those who are born of God must realize the importance and necessity of wholly abandoning the use of tobacco as well as alcohol, not only for their own good and safety, but on the higher ground of the good of others, which their Christian example will tend to promote. The use of tobacco, it is well known, blunts the moral sensibilities, dulls the intellect, and not infrequently ends in paralysis of the body.

"What right have we to wantonly impair, and even destroy, our moral, mental, and physical faculties, which were given to us to employ to the glory of God and for the good of our fellow-beings?

"If any professing Christian addicted to the use of tobacco believes that the evils and sin of this habit have been exaggerated, let him yield himself up wholly to the influence of the Divine Spirit, let him lay hold of the substance of Christ's teachings. and he will then perceive the gross selfishness of this demoralizing habit.

"I have been constrained to venture these convictions concerning the fatal tendencies of this all-prevailing vice, and the pernicious influence which professing Christians exert by the power of their example, by an earnest desire to impress them upon others as I believe the Lord Jesus Christ has impressed them upon my soul."

A FAITHFUL OFFICIAL.

I seize this opportunity to publicly testify my appreciation of the services rendered The New York Christian Home by Mr. E. M. Hayes, in his official position, and many a heart in this Institution will join me in praising him for his untiring efforts to make each one feel that when entering here he is indeed "coming home." Courteous to all, his kind treatment of each person endears him to those with whom he is brought in contact, and of him I may truly say that the milk of Christian politeness is the cherished nourishment of a soul rejoicing in the urbanities of the Lord.

GOD'S KEEPING POWER.

Approaching the end of this labor of love for Christ's sake, I feel myself impelled to say a few words in behalf of God's keeping power. One of the most difficult things to have fully realized by even some professing Christians in connection with the work of this Home is the fact that so many of the men here lifted from "out of a horrible pit, out of the miry clay," remain steadfast. These good people, who would feel as if highly insulted were they called "doubting Thomases," by the very fact that it is difficult to have them believe in the permanent character of the reformation here effected, imply a doubt in the efficacy of God's keeping power. I have had even a minister of the Gospel, a doctor of divinity, so utterly confounded at a visit he made here by the earnest, reverential attitude of the men in one of our meetings, that he asked, " What is it keeps these men ?" and was dazed, as it were, when I quietly replied, " The grace of God." But is not His grace all sufficient? Did not Jesus come to save sinners? and in the category of sins drunkenness is no more specified or singled out than adultery and murder. Hence it is clear that as "he that heareth My word, and believeth on Him that sent Me, hath everlasting life, and shall not come unto judgment, but is passed from death unto life," it is as reasonable to assume salvation for the drunkard who repents and believes as it is for any other sinner who is prompted to go "and do likewise." It is true that the determination of Satan to tempt is in no wise abridged by the resolution of the Christian to pray that the will of the Father may be done on earth as it is in heaven; but the ability of the devil to effect his fell purposes is limited by that immutable pledge of the everlasting God, narrated by the Apostle Paul in the 13th verse of the 10th chapter of his first epistle to the Corinthians: " There hath no temptation taken you but such as is common to man ; but God is faithful, who will not suffer you to be tempted above that ye are able, but will with the temptation also make a way to escape, that ye may be able to bear it."

" Him that cometh to Me I will in no wise cast out," and " Though your sins be as scarlet, they shall be white as snow," and " I will never leave thee nor forsake thee"—these are promises of God the fidelity of which cannot be doubted by those who see in Jesus "the light of the world," who recognize in Him "a friend that sticketh closer than a brother," who acknowledge that He is the substitute "who His own self bare our sins.'

The unfortunate drunkard says, "I will arise, and go to my father." He goes to the Father through Jesus. He has the promise of abundant pardon. He looks into the face of God and cries, "Though he slay me, yet will I trust Him." He learns that "The Lord is nigh unto them that are of a broken heart, and saveth such as be of a contrite spirit." He hears the voice of God affirming that "Whosoever shall call upon the name of the Lord shall be saved." He calls upon the Lord. The precious gift of saving faith is given him, and, no longer relying on himself or any human person or human thing, he leaps where "underneath are the everlasting arms," exultingly singing hosannas to the Son of David, knowing at last that justification has come "by His grace, through the redemption that is in Christ Jesus."

Dare any one say that when a poor drunkard thus throws himself upon God that salvation is not at hand ! No ; the man who thus seeks release from alcohol or any other besetting sin knows that he is saved. He knows it because "he that believeth on the Son of God hath the witness in himself" that he has entered among those who "are sanctified through the offering of the body of Jesus Christ once for all." He knows it because as "by one offering he hath perfected forever them that are sanctified," so "this is the covenant that I will make with them after those days, saith the Lord ; I will put my laws into their hearts, and in their minds will I write them ; and their sins and iniquities will I remember no more." He knows it as I know it, by personal experience and daily observation. He knows it because he fully realizes that "If any man be in Christ, he is a new creature ; old things are passed away, behold all things are become new," and while the salvation which Christ has *worked in* the regenerated man *works out* he goes on his way rejoicing, free in "the glorious Gospel of the blessed God," crying out with the sweet singer of Israel, "I will sing praises unto my God while I have my being," because, "The Lord is my defence, and rock of my refuge."

Yes, as the man, once a wretched slave of alcohol, opium, morphine or cocaine, but now a child of God, goes forth, hearkening to the voice of the Spirit which whispers, "Fear thou not, for I am with thee ; be not dismayed, for I am thy God : I will strengthen thee; yea, I will help thee, yea, I will uphold thee with the right hand of My righteousness," he sings these words, which consecrate his will forever :

> Laid on Thine altar, O my Lord divine,
>   Accept my gift this day for Jesus' sake ;
> I have no jewels to adorn Thy shrine,
>   Nor any world-famed sacrifice to make.
> But here I bring within my trembling hand
>   This will of mine—a thing that seemeth small,
> And only Thou, dear Lord, can'st understand
>   How when I yield Thee this, I yield mine all.
> Hidden therein, Thy searching eye can see
>   Struggles of passion, visions of delight,
> All that I love, or am, or fain would be—
>   Deep loves, fond hopes, longings indefinite.
> It hath been wet with tears and dimmed with sighs,
>   Clenched in my grasp till beauty it hath none.
> Now from Thy footstool, where it vanquished lies,
>   The prayer ascendeth, may Thy will be done.
> Take it, O Father, ere my courage fail,
>   And merge it so in Thine own will, that e'en
> If in some desperate hour my cries prevail,
>   And Thou give back my gift, it may have been
> So changed, so purified, so fair have grown,
>   So one with Thee, so filled with peace divine,
> I may not know or feel it as mine own,
>   But gaining back my will may find it *Thine*.

Yes ! God's keeping power is omnipotent as His saving power, and from out this Home we have sent hundreds of men who, as they are and prove themselves, witnesses of His word in time will be confessed by Jesus before the Father's throne and enjoy all the unspeakable, inconceivable beauties of the Beatific vision for all eternity.

VISIT US AND SEE.

A visit to the Home will give you a true insight to all its workings. As before stated, we hold services on Saturday evening, to which the public are cordially invited. Our statistical tables will be interesting to the inquirer, and will be found to be simply wonderful in their conclusions. Our meetings are continued without interruption all through the warm weather. From the outset the work has been a constant surprise to those interested, not only as regards its magnitude, but also as regards the readiness with which these men are brought under Christian influences and led to a better life. It is not possible to tell all the story of the Home, but it has more than justified the wisdom of those who organized it, and been the means of bringing joy and gladness to many sad hearts. From June 7th, 1877, to the present time, June 7th, 1887, it has cared for 2,222 men, not moderate drinkers, but drunkards and other intemperate men in bondage to appetites no less vile, such as morphine or cocaine. Many of these men were needy, destitute, friendless, and of them the great majority are to-day saved men, supporting themselves and their families.

The Home has thus far only in part been self-supporting; those of the inmates who are pecuniarily able have paid a reasonable amount for board. The deficiencies have been made up by those earnest friends who have had faith in the object and methods of the Institution.

PARTING WORDS.

In bidding adieu to those who have been, I trust, not wholly uninterested readers of the foregoing pages, I desire to draw special attention to the appended statistical tables, letters from former members of the Home, anti-tobacco items, bequests and funds, obituary notices, newspaper notices and other supplementary matter, all of which serve to strengthen the views herein advanced and cause hearts to glow with gratitude to God, to Whom for all His unspeakable gifts to us be praise, honor, and glory, given now and forever. Amen.

SUPPLEMENTARY MATTER.

STATISTICAL TABLES.

From June 7th, 1877, to June 7th, 1887.

Whole number of men received..2,222

NATIONALITIES.		OCCUPATIONS.	
United States	1,602	Actors, 14; Army officers, 6	20
Great Britain	474	Architects, 1; Book-keepers, 147	148
France	7	Brokers, 63; Civil engineers, 24	87
Germany	45	Clerks and salesmen	669
Canada	69	Clergymen, 18; Druggists, 31	49
Asia	3	Deputy Sheriffs, 3; Engravers, 15	18
Holland	2	Farmers, 3; Hotel proprietors, 17	20
Switzerland	4	Journalists, 48; Lawyers, 85	133
Denmark	2	Liquor dealers	8
South America	2	Manufacturers	49
Greece	1	Mechanics and trades	398
Sweden	6	Merchants	175
West Indies	3	Miscellaneous	202
Cuba	1	Musicians, 5; Naval officers, 14	19
Porto Rico	1	Physicians	65
		Printers	88
		Policemen	6
		R. C. Priests	7
		Soldiers	3
		Sailors	20
		Surveyors	1
		Teachers	22
		Telegraph operators,	15
	2,222		2,222
Pay members	785	Single	1,101
Part pay members	92	Married	708
Free members	1,345	Separated	*413
	2,222		2,222

Of these 2,222 men no less than 1,878 professed to be converted, while, so far as can be ascertained at present writing, 1,434 have remained steadfast.

Average time of drinking, 18 years, 9 months. Average time of excessive drinking, 8 years, 10 months.

215 chewed tobacco, 578 smoked, and 1,043 both chewed and smoked.

1760 obtained situations.

From the record kept from the year 1880 to 1886 respectively, we find that 1142 had either a Christian father or mother, and in many instances both parents were Christians.

Only one-sixth had intemperate parents, thus showing the fallacy of inherited appetite.

Two-thirds claimed that associations were the cause of their drinking.

* As near as we have been able to ascertain, 200 of these families have been reunited.

From the mass of correspondence received here from those who were redeemed, regenerated, and disenthralled in our Home, renewed in their manhood, emancipated from the bondage of alcohol, opium, morphine, and tobacco, and planted firmly on the truly good soil of Christ's all-embracing salvation, we select the following, written at different times and in various places, but all confessing the glory of Jesus, mighty to save, and thanking The New York Christian Home for its labors as the zealous instrument in God's hand to fill hearts and homes with that peace which passeth all understanding :

This man was for many years an importer of diamonds and watches in this city. One of our first merchants. Married an estimable lady. Lived in affluence. When he came to me was the picture of despair. No home, no friends. Deserted by all. He was saved in this Home. Came to me June 7, 1877.

MYSTIC BRIDGE, CT., Sept., 3, 1877.

My Dear Friend:

I have been intending to write you every day since the receipt of your kind favor of 21st ult., but now I can inform you that I have commenced in an open manner to work for our Lord and Master. My pastor has called on me, and I showed him your letters and also told him of the great work you were engaged in. I spoke in last Sabbath praise-meeting, as also the week previous, and have entered the Bible-class of the Sunday-school. I also attended the Thursday evening prayer-meeting at a private house, so that now I have got fairly started I shall go on all right. To put myself in proper manner before the old Christians here I had to tell my story or I should have felt as if I was deceiving them. It did not seem honest for me not to do it, and now I feel better and am glad I did so. I forgot to say that I attended the Y. M. C. A. meetings, and spoke there, and shall continue to speak for our Saviour, who saves to the uttermost all who will believe on Him. I took up smoking for a while after coming here, but I have absolutely left tobacco for good this time. I know it is a great risk to us and that it is wrong for me to smoke. I have stopped it now for all time to come.

It was pleasant for me to read of the Home that all have been steadfast. I pray every day that God will make them strong and enable them to remain true to the end. Your friend in Christ, *

A victim of despair and wretchedness. Left to die in the streets of our city. A good Samaritan recommended him here. Left us a true Christian. Came to us September 3, 1878.

BOSTON, October 11, 1878.

Do not think I have forgotten you, or the dear Christian Home I pray for daily.

Since I left you, God has given me grace to withstand all temptations, and with truth I can say I have not tasted liquor or tobacco in any form. I ask an interest in your prayers that I may grow stronger day by day.

Please remember me most kindly to your dear wife: also Brothers T—— and I——.

To-day I secured a position with a firm to travel South and West, a trial trip of three months. Yours, most truly,   *

A noble young man. From a western city of this State. "I came to be saved," as he said. All were in love with him for his noble Christian example while here. Came to us May 24, 1878.

E————, N. Y., October 19, 1878.

DEAR SIR :—I am not sure that I have a really good reason for not writing you before, so I will not bring condemnation on myself by attempting to frame an excuse. I have not refrained from writing you because absence and distance had begotten indifference, for I am sure not a day has passed that I have not thought of you with the keenest sense of gratitude. Not a day has passed that I have not regularly remembered the Home and its beloved Manager in my hours of prayer. I could hardly express the deep sense of gratefulness that fills my soul toward you as I remember how plainly you led me to that Christ who has since been so precious to my soul. What should I do without the presence, comfort and guidance of that dear Saviour? His Spirit has so dwelt in my soul since I left the sacred influence of the Home that even the thought of faltering has not occurred to me. My hours of private devotion are most precious seasons to me. There I lay in my supplies of strength, and there the Spirit witnesseth with my spirit that I am born of God. As I was in the habit of saying to the friends of the Home, I now say, "I'm on the rock." Truly my goings are established. I am trusting in my Saviour. He comforts me daily. The yoke is easy and the burden is light. How shall I ever be thankful enough to that God whose loving Spirit led my steps to your Home!

How often I pray God's blessing on you, dear Brother Bunting, and on your grand mission. You being my spiritual father, it seems as if my life now dates from my entrance at the Home. A mighty burden of sin and guilt has been rolled away, and a precious rest and light have flowed in and filled my heart. All things have appeared so bright since I have been home. I have been received with such tokens of confidence and affection by my friends and acquaintances here that it is no wonder things have a brighter coloring for me. For all this I give praise to God. He has enabled me so far to discharge every duty that has been made plain to me, and by His precious grace I expect to remain victor to the end. He has answered my prayers since I have been home in almost a miraculous manner. In truth I am trying to dedicate all my gifts and possibilities. all I have and am, to the service of the Lord Jesus Christ—to take all good things He may send me with grateful humility, and all chastenings which His love may impose with the same spirit. I rejoice in His love this morning. He whispers peace in my soul. I remember and pray for your meeting this Saturday evening. God bless the young men of the Home! And God bless you, my beloved friend and brother in Christ, with a most precious outpouring of His grace and comfort into your soul.

I know how anxious and troubled you were at times. I pray that, as these responsibilities weigh upon you, your soul may be most sweetly sustained by the precious consolations which Christ alone can bring to His children.

Give my love to all.   Yours affectionately,   *

This man came to me a perfect scoffer. Had not been in a church for over twenty years. Apparently a hopeless case. The Holy Spirit softened his heart. He became a true and devoted Christian. Came to us in October, 1879.

N. Y., December 5, 1879.
My Dear Brother:

Your very welcome letter was duly received; it brought me great joy. To hear from *you all* is truly a pleasure. I received one from Brother B., who said that you had a good time Thanksgiving Day, though I would like to have been present. Do you know that it is just two months ago to-day that I first made my appearance at the Christian Home—*God's Home—my Home*—and I hope that the day is not far distant when I shall make my reappearance; you do not know how much I miss *our* Wednesday and Saturday evening reunions. I hope that you will remember me to them all, for God knows that I remember you all every night and morning in my prayers. As there are those in the Home I have never known personally, tell them of my case, assure them that it is only with the love of our Saviour that they can be saved. If they will put their trust in Jesus Christ He will never forsake them, for He has not forgotten *even me*, the poor miserable sinner that I was. No: He died to save us all. Have faith, brothers, and you will all surely be saved. One night this week two of my old acquaintances called upon me at my hotel and sent up their cards. I went to the office and sat down to chat with them. The first thing was, "Come, let's have a drink." I said, "I have taken my last drink some time ago." "Well, then have a cigar." "I neither smoke nor drink." Well, *they* were disappointed. *I* was delighted. I still continue to attend the Y. M. C. A. They have very interesting meetings. I wish you would ask my good Brother McB. of the Y. M. C. A., who directed my wife to the Christian Home, if he will send me a letter of introduction to the Secretary of the Y. M. C. A. here. Please give him the heartfelt thanks of myself and wife, also to Mrs. B. and all. Write soon. Yours, &c., *

A godless man, a scoffer. His wife came to the Home and remained a week with him. At one of our morning Bible readings the Holy Spirit gained admittance to his heart. I was in conversation with him over two hours. Had not read a word in the Bible for twenty-seven years. As soon as he was saved he rushed out of my room to his own, and there on his knees begged for salvation for his wife. Died in full trust of God's promises. Came to us October 5, 1879.

—— —— Ky., Nov. 23, 1880.

My Dear Mr. Bunting:

Your very welcome letter was forwarded to me from S——. My address is as above, and has been since last Spring. Well, you ask how I am getting along in the good way. I can assure you, first-rate; since I entered the *Christian Home* I have not used a profane word. I have said my *prayers* night and morn, never once forgetting the *Home* and all its inmates. Neither have I drank a drop of anything stronger than tea or coffee. Neither have I used tobacco in any form. I wish J. S—— could say the same, but my opinion is that he does it more for an advertisement than for the love of it. The man you spoke of, J. P——, was never in my company, but with my brother. I am very glad to learn that he has reformed. I sent you a paper the other day, giving you an account of my business. I am doing exceedingly well. So you can see what *letting rum* alone has done for me. Just say to my unfortunate brothers to do as I have done—let rum alone—lead a good Christian life—go to church on Sundays—and *God* will do the rest. Please remember me to each and every one; tell them that I think of them and the *Home* often. This indeed ought to be a *Thanksgiving week* with me, and my friends also. I shall be pleased to hear from you soon. My wife is here with me, and together with myself, send our kind regards to Mrs. B——.

Sincerely yours, —— ——

I met this man in one of our Gospel meetings. He was all gone. Had given up all hope. Wanted to be saved if the day of redemption had not passed. Was in earnest. Asked God to hear his prayer, "God be merciful to me a sinner." It was answered. He is a wonderful work of grace. Came to me November 14, 1880.

NEW YORK, August 17, 1881.

Dear Mr. Bunting:

I thought I would give you another letter as I promised. I have nothing new to write about, but an interchange of thoughts and feelings sometimes brings great good to our hearts. Everything moves along very pleasantly, and the good old promise is fulfilled day by day, that all things work together for good to them that *love God and are called according to His purpose.*

I must say, Jesus and His salvation never were more precious than now. I behold Him in all His beauty and power in my own salvation. Ah, think what boundless love and what matchless grace could save a wretch like me! and I am saved, and I know it, blessed be God the Father, the Son, and the Holy Ghost. It took the entire Godhead to do it, but it is an entire and complete salvation. Who wants anything better? I don't: it is for time and eternity. What more can I ask? Nothing. Glory to the Triune God. I have life more abundantly! full of joy, peace, and perfect satisfaction, and hereafter everlasting bliss, companionship of Abraham, Isaac, and Jacob, and all the heavenly host forever and forever.

O my dear Brother, exhort everybody to come and join this holy band, and on to glory go. Tell them how God saved me, even me. When friends forsook me, society hurled me out, relations disowned me, and all, with one consent, put aside all efforts to save, and concluded I was past hope or redemption, and one united shout went forth, "Let him alone." O blessed Jesus, O blessed Holy Spirit that directed, Oh blessed Christian Home, Oh blessed man of God, that pointed me to the Lamb of God! Oh, my heart is bursting with joy and praise; Glory to God forever and ever. Give my love to all.

Yours affectionately, —— ——.

This was indeed a wonderful salvation. He was in despair. A perfect wreck, physically and morally. His wife, as you see, was in despair. She was a R. C. Through his conversion his wife was saved. His testimony is heard often in our Saturday evening meetings—the letter-carrier—Every one rejoices when he testifies. Came to us November 2, 1831.

NEW YORK, November 2, 1882.

Mr. Chas. A. Bunting:

DEAR SIR :—It is with pleasure that I write this letter to you. One year ago to-day my husband and I went to your "Christian Home" in Seventy-eighth Street; when I left him that day I was very discouraged, for I had no hope it would do any good. He had tried so many times to stop drinking before and failed. He told me himself, after being there three days, it was no use, he could not do without drinking rum. I left him that evening, feeling worse than ever. Before I had seen him again he was changed so much that I could not believe he was telling me the truth, because I was sure that nothing but a miracle, done by God, could save him. He told me then that he did not know himself how it was; but that, before going to bed the night before, he had asked God very earnestly to take the desire away from him, and in less than a minute he felt sure he was answered. He went to bed, and slept well for the first time in months; and getting up the next morning, he was surprised to find the desire and appetite all gone. I think it was just as great a miracle as I ever read of in the Bible. He has not only been saved from rum, but he has become a good man every way. I only wish I was as good. And our boy has also been spared to us through the prayers offered up in your blessed Home for him. I feel sure of this, because eight of our best doctors in this city had given him up, and said he could not be cured: so you may know, Mr. Bunting, that it is with heartfelt gratitude to you for your interest in him that I write this to you. We have troubles, of course, for I have been sick a great deal lately, but it makes no difference to my husband; he is better and more steadfast every day.

I thank God that he put it into the hearts of those gentlemen to build such a Home, for they have made many wives' and mothers' hearts happy. I hope by the time you receive this you will have recovered from your illness.

I also thank Mr. H. for his many encouraging words to me and my husband, and also Mr. Pulis. Yours respectfully. —— ——.

A gentleman by birth and education. A lawyer by profession. Had not drawn a sober breath for six months previous to his entering our Home. Gave himself without reserve to Jesus. Was saved, and from the day he arrived in his own city has been a worker in Christ's vineyard. Came to us Dec. 31, 1882.

WASHINGTON, D, C.. Sept. 18th, 1883.

DEAR BRO. BUNTING :— . . . Oh! how I praise and adore the blessed Lord for what He has done for a poor sinner like me. Dear Bro. Bunting, there is not a day of my life that I do not take from my pocket-book the precious *talisman*, the ever-present pasteboard you gave me at the Home, and read those blessed words: "When tempted or tried, either in prosperity or adversity, *honor God by declaring at once.* ' *Jesus saves me now,*' " and to this may I not, in all truth and soberness, ascribe my unyielding faith in the precious Master and my steadfast adherence to a Christian life? I have no love for the world *per se*, for there is nothing in it (I have tried it to the uttermost) that can give absolute satisfaction to the weary soul of man. I never again desire to love the world.

> Its dreams, its songs, its lies :
> They who have followed in its train are not
>   The true, and good, and wise.
>   The wise and good,
> They choose the better part :
> To the true world that is to come they give
>   The true and single heart.

I am striving the best I know how to set my affections on things above where Jesus sitteth at the right hand of God, to make intercession for me and to plead my cause at the bar of heaven. In the case of our fallen friend whose name I have written sorrowfully we are forcibly reminded that it is too true that "No man can serve two masters," for "whosoever is the friend of the world is the enemy of God." In conclusion I will add that I am laboring to the best of the powers with which I have been endowed by my Creator to work in the vineyard of the Lord, and have chosen as my field the grand and noble cause of Gospel Temperance, and as long as my strength endures and health continues by the gracious favor and support of Almighty God, it is my unalterable purpose to do battle in that holy cause. and may God speed the work and enable me to do something toward the reforming, the saving, and keeping reformed and saved precious souls who stand so much in need thereof. Your weak but earnest Brother in Christ, ——— ———.

This man was brought here in the morning after sleeping in the park. A thorough gentleman. By education qualified for any position. Became a minister of the Gospel, and is now doing well. Came to us September 26, 1882.

——— ———. Mass., March 14, 1884.

My Dear Mr. Bunting:— ✻ ✻ ✻ ✻ ✻ ✻ ✻ ✻ ✻

In my judgment you give no instruction in your daily services that is more important than the plain, practical and earnest lessons on the use of tobacco. It is true, members frequently complain of your constant warning against its use after leaving the Home. They seem to think that they know better than you; that their temperament or constitution are different from those who have made the same trial and miserably failed. *After one day's use of tobacco*, 1 am alarmed, sad and trembling as to the future. This few hour's use of the poison has again aroused all of the old gnawing and thirstings for drink. God alone can prevent this horrible sequel, and to Him must I flee for succor. In my judgment, tobacco is the most powerful yet secret agent of the devil in leading one *back* to the intoxicating drink. I care not whether it is the *tobacco itself*, or "letting down the bars," one thing is certain, it is a compromise. Manhood, temperance, a strictly honorable life with God, are all compromised when we again use that from which we once broke loose and admitted to be treacherous.

I write this from a sad experience and to warn in a kind, Christian but decided manner those who are now in the Christian Home, and who do not fully and conscientiously agree to these thoughts so rapidly and briefly suggested.

I am, with great respect, your obedient servant, ——— ———.

OPIUM AND ALCOHOL.—In despair, with no hope, this man came to us one afternoon with his nephew He was filled with opium. He denied having taken any that day. I afterward found secreted about one-half an ounce in his clothes. Read his own story. Came to us March 6, 1884.

CONN., June 29, 1884.

Dear Mr. Bunting:

As you know, I became an inmate of The New York Christian Home three months ago, and when I entered it I was a moral and physical wreck. I was a slave to the use of opium, and that, together with *rum* and tobacco, had undermined my health and left my reason waning in the balance. So bad had I become that my case by many was considered hopeless. I cannot recall any perceptible improvement in my case until I had been there nearly two weeks. I was called into the Manager's room and there was taught the true way to be cured. The remedy is simply Jesus Christ and Him crucified and risen. The teachings were presented to my mind so plainly that I could not help understanding them.

I shall ever bless God that I was led to this Home, for I feel confident that I could be saved in no other way. I had tried all earthly remedies, but of no avail. My friends now, as I meet them, stop me and ask, "Is it possible this is you?" They see so much improvement in me that it does not seem possible that it can be the same man they saw six months ago. Well, I tell them they must give God the glory and the Christian Home the credit. I cannot refrain from saying a word in regard to the management of the Home. I am convinced that in no other place (known now to me) could a class of men such as are found there be cured of their malady, as in no other place would they get the teaching as we get i: there. We listen day after day to Bible lessons by the Manager, and he always has something new, and not only that, but it is given to us in so plain a manner that the simplest can understand and the wisest cannot refute. I do think this is not fully understood by those outside the Home, and feel confident that when this is rightly understood the Home will not be large enough to hold those that will seek an entrance through its door. May God bless and prosper the Home is my prayer. I am to-day a monument of its fruits and blessings. Freed from the curse of opium and drink, reason restored and health good, can I ever bless God enough for my freedom? I would say to all who become members of the Home, do as I did, and never leave there until you are sure Jesus has forgiven you, and that you have consecrated yourself fully to him. I miss the Home and its influence and teachings, but I can truly say, though away from its influences, I realize that Jesus is able and willing to keep me. How often I repeat that assuring motto taught at the Home, that when we are tempted and tried let us be able, by strict obedience to God's commands, to say, "Jesus saves me now." May God so direct me in my future life that all I may do may be to His honor and glory, for I am sure that so long as I profit by the lessons taught me in my spiritual home so long will I be in the path of safety. Yours, M. K.

This son was given up by almost every one. His brother had become so disgusted with him that he wrote me he never wanted to hear from him again. He is now an evangelist in a neighboring city. Saved by the power of God. Came to us September 21, 1883.

(*From the Mother and Sisters of a former Inmate.*)

LEGHORN, ITALY, Nov. 29, 1884.
3 Via Degli Elisé.

Mr. Charles A. Bunting, Christian Home, Madison Ave., N. Y.

DEAR SIR :—It is now about a year since my son was an inmate of the Christian Home. We thank God, who directed him there, and we thank you and those associated with you, for all the kindness he received from you and them. We are grateful beyond expression for the change in him. He has told us much about the Home, and how many happy hours he has spent there ; and we have shed many tears of joy over his conversion. To God be all the glory.

My daughters and I have sent by P. O. order 105 francs, which I think is equivalent to about twenty dollars. It is a very small token of gratitude and good wishes. Believe me, Yours truly, —— ——.

Came to us, as he remarked, to be saved: His brother was then a member of the Home. Soon he gave himself to Jesus, and from that day to this has never faltered. Came to us January 15, 1884.

————, CANADA, Feb. 3, 1885.

Dear Mr. Bunting :

You must excuse me for not writing sooner, and I now take the opportunity of sending you my hearty and sincere thanks for all your love and kindness to me in bringing me the dear and loving Saviour, who loves me dearly, and "saves me now ;" and God be praised for His gift in sending His Son into the world to save such a poor, weak sinner as I was.

Since my return I have not had the slightest desire for tobacco or strong drink, and my whole house is now so sweet and nice since giving up the narcotic weed.

Give my kindest regards to Mr. Pulis, my brother, and all the dear members of The New York Christian Home, the only Christian—*i. e.*, the best—home I have met with in my travels—the home where I found *peace* and *rest* for my immortal being.

James says in his 5th chapter, 15th verse, "Pray one for another." *Pray for me*, that I may be kept steadfast in Him who has indeed made me free. Remaining, dear Mr. Bunting, Yours in Christ, A.

Opium and rum—a slave to both habits. A wonderful victory. A gentleman of education, but near his end when he came to us January 15, 1884.

My Dear Mr. Bunting:

———, CONNECTICUT, December 20, 1886.

Nearly three years have now passed away since I became free from the use of opium, and when I consider what I have been relieved from, I can say continually, " God bless the Christian Home." When I came to you three years ago I was considered a hopeless case. I had become a slave to that worst of habits, opium. This is a thousand times worse than rum. My system, physically and mentally, was a wreck, so much so that for three weeks after entering your place I was considered hopeless by all except yourself. You had faith in the Lord's healing power, and stood by me. In two months' time I left the Home, strong in body and mind. I have never, from that time to this, tasted of opium nor alcohol of any kind, nor have I had a desire to do so. I send you these words so that you may feel encouraged in your noble work, and not be afraid to try any case, however dark it may look. Truly and justly have you been called by God for this great work, and may you long be spared as the instrument in God's hands of leading others into the same new life as you did me.

Yours truly, ———.

Another victim of opium and whiskey. Came to us all broken up. A perfect picture of despair. Became one of the most earnest of Christians I have ever known. Makes every one happy when he testifies in our meetings. Now occupying a trustworthy position in one of our city banks. Came to us May 24, 1879.

NEW YORK, May 24, 1886.

Dear Mr. Bunting:

I am happy beyond measure to say that on the 24th day of May, 1886, I entered upon the eighth year of my Christian life, having conquered the opium habit and found Christ on that date in the year 1879, in your little study at the Christian Home in 78th Street. To-day, *now*, I am still clinging to the Rock, and Christ, as my constant guide, comfort, and support, has become a blessed reality. His word is a lamp unto my feet and a light unto my path, and day by day the good Lord becomes more precious to me, not alone for what he has done for me, but also for what he is continually doing for me. I just give myself away to Jesus every morning, and I find that he smooths my pathway, and with my heart full of the love of Christ I cheerfully follow wherever he leads me.

I wrote to you on the 24th of last May my usual letter, for you know as long as you and I live, *that day* shall never pass without my annual letter to you.

My dear Mr. Bunting, I have you in my thoughts probably more than you think, for you led me to Jesus, and that kind act of yours, and the way you have encouraged me, makes me your everlasting debtor. God bless you, my dear friend, and may you be a blessing to many, *many* more souls, in leading them to the Saviour, for you were the instrument, in God's hands, that gave me peace, joy, happiness, and everlasting life. Pray for me, and think of me occasionally. I will be up soon to see you.

Ever your sincere and grateful friend, ———.

HOPE FOR THE SLAVE OF THE CUP.

Large brained, large hearted, generous men, permeated with a love for the fallen and afflicted, have given freely of their abundance to found and place on a permanent basis an institution for the salvation of those who have lost their self-control, by affording them opportunities to become regenerated men.

The New York Christian Home for Intemperate Men is located at No. 1175 Madison Avenue, corner of 86th Street, on one of the finest and healthiest sites in New York, with Central Park only a block distant. The arrangements and management of the Home have been perfected, after years of intelligent study, by Manager Charles A. Bunting and the Directors of the Institution. The members of the Home are impressively reminded that drunkenness is a sin against God, to be repented of and forsaken. In fact, the chief and distinguishing advantage of the Home over other institutions is that from the first it environs every member with the persuasive, elevating atmosphere of simple Gospel truth, applied directly to the sin which has brought such misery into his life. The instructions and teachings that the members receive are singularly effective for the redemption of the inebriate. It was the good fortune of the writer of this to have enjoyed the invaluable blessings and advantages of the Home for several weeks, and he will never forget the ringing, heartfelt testimonies and experiences given on Saturday evening by the members and ex-members (many of them now scattered over the United States and Canada). They evidenced thorough reformation and regeneration. Their testimonies came from their hearts, and their countenances were full of joy as they acknowledged their thankfulness to God for their deliverance from a worse than Egyptian bondage. Such almost miraculous restorations extort the acclaim : " Verily, nothing is impossible with God !

The daily life at the Home is by no means monotonous. Religious services at at 9 A.M. and 8 o'clock P.M.; prayers at 9.30 P.M. Opportunities for walks in Central Park, the privileges of some of the best churches in New York on Sunday, and Wednesday and Friday evenings, daily papers, a well selected library, sympathetic company, and regular hours, regular, wholesome, generous meals—all combine to make the establishment a delightful Christian Home.

An exceptional feature in the Home is that Manager Bunting (a God-fearing, noble-minded Christian gentleman), who has dedicated his best talents and energies to his work, and his Assistant Managers, Messrs. Pulis and Hayes, and, in fact, all the officers in the Institution, are regenerated men, who in times past were under the bondage of the cup, and have been reclaimed by God's grace. Consequently all are bound together by the ties of human sympathy and Christian brotherhood. Grateful unto God for having been a partaker of its benefits, I fervently exclaim : "God bless the Christian Home !"

GEORGE J. BRYAN, in *Buffalo Christian Advocate*.

[Truly, the grace of God can reach all. Mr. Bryan came to us for help out of his terrible life of sin. He was a gentleman of culture and education, with a lovely family, all in distress. Once a leading editor in his city, but everything was gone. Since he left us he has regained his position in society and has his old paper back ; is now editor and proprietor, and is loved and respected by all who know him. Truly, God is good !

We publish this, as it will, no doubt, help some other poor, lost one to seek this haven of rest.—C. A. B.]

ANTI-TOBACCO ITEMS.

CIGAR stumps are gathered from the streets and gutters in many of our popular cities and sold to cigarette manufacturers, and again find their way into market and are then smoked by ladies and gentlemen. No accounting for taste.

EVEN a century ago, Dr. Rush discerned the closely connecting link between alcohol and tobacco. He said : "Smoking and chewing tobacco, by rendering water and simple liquors insipid to the taste, dispose very much to the st·onger stimulus of ardent spirits."

SPECIAL observations of the effects of tobacco on thirty-eight boys from 9 to 15 years old have been made recently by Dr. G. DeCaisne, a distinguished French physician. With twenty-two of the boys there was a distinct disturbance of the circulation with palpitation of the heart, deficiencies of digestion, sluggishness of the intellect and a craving for alcoholic stimulants. Twelve boys suffered from frequent bleeding of the nose, ten from agitated sleep, four had ulcerated mouths and one had contracted consumption, the effect of deterioration of the blood. produced by long and excessive use of tobacco.

"Why did you learn to smoke, my boy?" "For the same reason you did, I suppose." "Well I want you to stop smoking." "Won't you give me the reason for stopping that I had for learning, father." After a moment, "Yes, I will." Both stopped.

———

A zealous preacher, who loved smoking better than he ought, in a heated discourse, exclaimed, aiming his rifle at some of his hearers: "Brethern, there is no *sleeping car* on the train to glory." One of the party whom he aimed to hit responded: "No, brother, nor smoking-car either."

———

When we see church members paying from six to ten dollars for tobacco, and only two to four for the Gospel per year, we are forced to conclude that if a man will rob God of His tithes and offerings from love of his pipe, it is high time to cast to the moles and the bats the idol that claims such a supremacy.

———

The *Arkansas Methodist*, thus speaks of the tobacco habit: "Bad looking sight! A woman learning her children to say their evening prayers with an old stick in her mouth full of snuff! A snuff-dipping mother, with a tobacco-chewing father, aided by a smoking pastor, with a plenty of tobacco, form a fine circle to talk Christian temperance, and teach the children self-denial."

———

"How careful ought every Christian to be not only before a criticizing world, but everywhere. A single word or act may unintentionally warp the mind of a weak Christian, as a breath of wind may sometimes warp the giant oak. Especially are ministers of the Gospel liable to do harm by their inconsistencies, and we notice right here a bad case, hoping it may have a good influence, not with a wish to do anybody harm. A layman proposed to a minister to quit chewing tobacco if he would quit smoking, but the minister declined."—*Arkansas Paper.*

I am advised that two laymen not five blocks from where I am writing offered to quit smoking and rent a pew in the neighboring church if the minister would quit using tobacco, and the reply was, "Show me anything in the Bible that says a minister shan't smoke and I will." No wonder the church lies dormant. Clergymen stand in the pulpit and exclaim, Why is religion so unpopular? Why so many vacant pews? Why so little interest in Christianity? and then in loud tones and long sentences they ask, "Is it on account of science falsely so called?" Is it on account of agnosticism? Is it on account of the teachings of Huxley, Tindall, Ingersoll, and the like? Oh no, dear friends, it is on account of not doing what the blessed Lord tells you to do, viz., "*It is good neither to eat flesh, nor to drink wine, nor anything whereby thy brother stumbleth, or is offended or is made weak.*" This is what's the matter! And a Methodist editor says:

"How can a preacher chew and smoke and come to conference whining over assessments not half paid : his own salary meagre, perhaps in debt, and troubled by day and by night as to how he is to come out: yet he and his people have chewed, smoked and dipped up enough money to have made him independent, and not one word has he said to them to stop this unwholesome, this fearful practice. No, he could not. for he was a partaker and encouraged it by his own valuable example. Oh, brethern, how will you meet this at the great day!

REV. DR. T. DE WITT TALMAGE. in a recent stirring and able sermon. said : One reason why there are so many victims of this habit is because there are so many ministers of religion who smoke and chew. They smoke until they get the bronchitis, and the dear people have to pay their expenses to Europe. They smoke until the nervous system breaks down. They smoke themselves to death. I could name three eminent clergymen who died of cancer in the mouth. and in every case the physician said it was tobacco. There has been many a clergyman whose tombstone was all covered up with eulogy which ought to have had the honest epitaph: "Killed by too much Cavendish."

WHEN the charter of our New York Medical College was granted in 1866. writes Mrs. Lozier. M. D., a medical gentleman and senator from the rural districts, who had favored the bill, sent me his congratulations, saying also that he had an only child, a daughter six years old, who, he hoped, when old enough. would become my pupil. About a year ago this daughter, now a young lady, was brought to me, not as a pupil, but a patient, her father reporting that she had always been too nervous to study, and that he could never trust her from his care.

Her symptoms led me to inquire concerning his habits in regard to the use of tobacco. He was an inveterate smoker, and because his wife found the smell of it unendurable, when in the house he confined his smoking to the study, where his daughter was his constant companion. The young lady's condition was critical; the action of the heart was so irregular that she could not lie down, and thus her sleep was interfered with. After I had seen her three times, and made an examination of her case, her father asked me, "What do you think is the cause of her illness?" "I am sure," I said, "that her condition is due to the inhalation of tobacco." After a little reflection he replied: "I believe it! Tobacco is an arterial sedative, affecting the whole circulation of the blood." Bringing the right hand down with decision, he exclaimed: "Mrs. Dr. Lozier, you have hit upon the right, I am convinced, and if I should ever take up a temperance crusade I would begin at tobacco." Notwithstanding that the invalid is somewhat improved since being removed from the poisonous atmosphere. I fear the truth is that her constitution is shattered for life.

FORM OF A BEQUEST.

I give and bequeath to "THE NEW YORK CHRISTIAN HOME FOR INTEMPERATE MEN," *the sum of*

....

and the receipt of the Treasurer thereof shall be a sufficient discharge of my executors for the same.

"He that giveth unto the poor shall not lack."

THE VANDERBILT FUND.

In connection with bequests to this Home, it is my pleasant duty to acknowledge the munificent legacy of $50,000, left us by the late William H. Vanderbilt, and which, invested with the knowledge and sanction of his executors in approved securities, gives the interest constituting the "Vanderbilt Fund." The income thus procured is a welcome and important addition to our resources, and stands as a permanent endowment for the relief of those who, through habits of intemperance, have been reduced to an utterly friendless and apparently hopeless condition. Only those engaged actively in the work of reclamation can tell the blessed benefits derivable from such a fund, because experience teaches us that from this very class of friendless men spring many who become not only thoroughly restored themselves, but leave here to specially and singularly help in reclaiming others.

The great interest taken in the work of The New York Christian Home by the late Mr. Vanderbilt was manifest in many ways. He had a keen appreciation of the blessings it brought to so many families and to the community. In every reclaimed man he saw a distinct gain, religiously, morally, and financially, to the Nation. He felt that however much of this world's goods a man might possess, he could not stand alone, and that in the aggregate of general profit and loss even the millionaire is affected by the failure, the ruin, the waste, the non-productiveness, which flow from habits of intemperance. With all the beneficent results brought about in the old Home he had such intelligent sympathy that at one time he seriously contemplated paying all the cost incident to the price of the lots for the new Home, and the erection of the building itself. Causes, unnecessary to detail, intervened between the generous proposal and its execution. But while he lived his donations were large and given with an abundance of spirit. Such an example need scarcely be urged upon the wealthy among Christian society as being worthy of imitation. It speaks for itself, and emphasizes the wisdom of following the Gospel injunction to lay up treasures where the rust consumes not nor the moth eats away.

In thus recording my sense of gratitude to the memory of William H. Vanderbilt, it is appropriate to mention that when he was called to bid adieu to earthly scenes, the friendly feeling which had been felt by his family as well as himself for this work did not slacken. On the contrary, his son, Cornelius, who had been a member of our Board of Directors almost from the very first, remains one of the most active and sympathetic among that body, was a large contributor to the Building Fund, and is still one of the largest contributors to the various funds needed to enable us to carry on our labors. Personally he has been more than kind to me, for when sickness, bodily infirmity and accidental loss superinduced monetary requirements beyond my means he, without a hint, much less a word of solicitation, so met my needs that it seemed as if he were receiving instead of conferring a favor.

THE MEMORIAL FUND.

Sometime prior to the date of the Tenth Anniversary of the Home it was deemed advisable to fittingly celebrate an event of such deep interest to all engaged in this mission of mercy, and to still further characterize our work as an eminently Christian endeavor, by starting a Memorial Fund of $5,000, to be applied in granting temporary loans to deserving indigent members on their departure from the Home, thereby enabling them to pay their board for a little time while seeking employment. It was felt that this would be a most suggestive way of aiding our poorer brethren, and as each loan was but for a time, the appearance of mere mendicity would be removed from the application of the fund, while the honorable return of each amount loaned would keep the gross total intact for the assistance of others. The celebration was held on June 7th, 1887, and was largely attended. Addresses were made by the Rev. J. M. King, D.D. (Methodist), Rev. D. C. Litchfield, D.D. (Baptist), Rev. George J. Mingins (Presbyterian), Caleb B. Knevals, Esq., on behalf of the Home, and Charles A. Bunting, Esq., Resident Manager. In his address Mr. Bunting reviewed the inception and progress of the work, detailing its early struggles and later successes, dwelling upon the kindness of those who had upheld his endeavors when the darkest clouds seemed gathering, all united in prayerful rest upon the heart of Jesus while urging the work forward like earnest laborers in the vineyard of the Lord. In closing his remarks, the Manager alluded in feeling terms to the many obligations the Home was under to Mr. Knevals, and also to Mr. Pulis, Assistant. A tribute to Mr. Hayes followed, in the course of which Mr. Bunting said that many a heart in the Home would join in words of praise to Mr. Hayes. The sum of over $1,900 was realized, and the Fund still remains open, to be added to by all who feel disposed now to attest their Christian charity for their needy brethren or their gratitude to God for the blessings they have reaped in, by and through this Home.

Obituary Notices.

H. B. T. came to us March 26th, 1879, and when he died several years afterward sent this message to me by Mr. Pulis: "I die trusting in Jesus."

Mr. Wiltbank came to the Home July 5, 1877. His indeed was a life of severe struggle. At last he died in the Roosevelt Hospital, and sent us this message by Mr. Pulis: "Tell Mr. Bunting I am saved."

Mr. Stanley was a very talented young man, and beloved by all who knew him. I had the privilege of standing by his bedside in the hospital. He died fully trusting in Jesus, and his funeral was attended by many of the members of the Home.

A. H. M., a noted minstrel, joined our Gospel Temperance band in the Home Oct. 5, 1879, and soon entered on a serious life as a Christian. Read his letter published elsewhere, dated from Kentucky, November 23, 1880. I have just learned through his wife, of his triumphant death.

Frank Budworth, the actor, died in New Orleans July 10, 1884. In this Home it was his blessed privilege to make open confession of his trust in the Saviour, and he lived steadfast and true to the lessons of faith and love, temperance and truth, until he fell asleep in Jesus, assured of a blessed immortality.

Mr. G., an actor, a lovely young man, came to us after trying many Reformatory Institutions without success. In this Home he found Christ to be his Saviour. He died in his mother's arms, and his last words were: "God bless the Christian Home. Tell Mr. Bunting I die trusting in Jesus." He was sick with consumption some eighteen months, and died two years after leaving the Home.

On the 19th of November, 1884, Mr. Oscar M. Wilcox, of this city, and formerly a member of this Home, died in Utica. N. Y. He was much beloved by all here, and it was at this Institution that he gave his heart to God, going forth from us a Christian man. He was a consistent follower of the Lord, happiest when doing good to others, and raising them up from the pit of destruction.

A Mr. Moore had remained with us for several months, being employed with duties in the Home. He was converted here, and all through his stay he lived a devoted Christian life. Shortly after leaving us he died of a very painful disease. It was the privilege of Christian friends to stand by his bedside, and as he passed away, to hear the words as he tried to sing, "What a friend we have in Jesus."

J. M. L., of this city, was received in the Home June 13, 1882, and here gave his heart to God. He left us on the 11th of the following month, and his uncle, Mr. Mackey, informed us that he continued to live a Christian life up to the time of his death, about two years after he had been saved. With his last breath he praised God for inspiring those who had brought him to the Christian Home.

———

Mr. M., of Baltimore, a very nice man, with the one exception of his besetting sin, became a Christian man while an inmate of this Home. During his sickness, he was in frequent communication with his wife and friends. His brother came on to this city expressly to take him home, but he was too feeble to be removed from the hospital. His last words to me were, "I am already to go. I am waiting and watching for Jesus."

———

Mr. P. became one of the Home family September 5, 1878, and here learned that Jesus could save to the uttermost. It was in connection with the efforts to inaugurate the Memorial Fund that we learned through his wife of his death, which took place July, 1884. She says: "As long as he was able to think or to speak, he had loving, grateful thoughts and words for the Christian Home and its workers, and for your kindness and interest in him." Read his beautiful letter from E——. October 19, 1878.

———

A Mr. J., of Providence, R. I., came to the Home when first opened. Soon after coming he gave his heart to God. He returned to his home. About two years after, he came to this city and called on us, at that time being very sick with consumption. These are his words: "Before going home I wanted to come and see you and your wife, and tell you that the teachings I had received while here were my salvation. I have made a somewhat crooked track, but my eyes have never been taken off of my blessed Saviour." He died in Providence, shortly afterward.

———

Firm in his religious trust, and true to the temperance vows he had made in the Christian Home, in the month of June, 1885, David Taggart, of this city, went to enjoy his heavenly rest. His was a Christian soul, full of love, hope, faith, and to the very last texts of Scripture and the words of familiar hymns lingered on his lips. While sick his physician proposed to give him stimulants, but he said, "No; I will never take anything of that kind again. I want to die sober." And sober he died to earth, only to live anew in heaven with the Christ who had touched his heart and given him the new birth in the gospel of everlasting life.

William S. Stroud became acquainted with me and the Home through the instrumentality of Rev. S. H. Tyng, Jr., in the month of January, 1881, and he gave himself to Christ without reserve shortly after. From time to time his testimony for Jesus was heard in our meetings. In the Fall of 1883 he was attacked by consumption, when, yielding to the earnest solicitations of my wife and myself, he came to us, that better care and attention might be given his then enfeebled body. It became apparent very soon after he became a member of the family that his stay with us would be of short duration. Toward the end of January, 1884, the physician in charge thought it best to let him know that in a few weeks the end would come. He had cherished a strong desire to see his mother in England before he died, and at his request I cabled to her. In her reply she said that if the doctor thought her son would live long enough to reach his native shores to let him come. On the 31st of January he sailed for his home, but died five days after, the immediate cause of death being the rupture of a blood vessel, superinduced by a severe coughing spell. His last words were, "Give my best love to my dear mother and to all." Thus he passed away, but we know it was well with his soul. Redeemed with the blood of Christ, he had fully surrendered himself to his Saviour, and his parting words to us are worthy of continued repetition—"Thy will, O God, be done."

On the evening of February 3, 1886, the angel of death visited the Home, and in the form of heart disease took from us our dear brother, Samuel Crothers. He had been a member of the Home but a short time, long enough, however, to grasp the great truth that Jesus died to save him. On the evening previous to his death he gave his first testimony for Christ in these words : " In the providence of God my feet were directed to this Home, and, oh ! how I do thank him for it. I have this day given myself to Jesus. He has saved me. Pray for me, brethren, that I may always be faithful." He retired to his room in his usual health, was in conversation with those about him, when he suddenly arose, dressed himself, and went down to the office and desired the one in charge to let him out, as he was very sick and wished to go home to his mother. He was prevailed upon to return to his room, and the doctor at the Home was summoned to his bedside, and while writing a prescription he passed away. A few moments before his death he wished the following message to be delivered : " Tell mother it is all right with me, I die trusting in Jesus." This is the third death that has occurred in the Home during the ten years of its existence. The first was a Mr. Quinn, who, while apparently in good health, died suddenly of apoplexy, with hardly a moment's warning. The second, Mr. Raymond James, ate his breakfast, but not feeling very well, returned to his room. Before retiring he read a chapter in the Bible and knelt in prayer. This was about 10 o'clock P. M. About three hours afterward he awoke the friend sleeping near him by his peculiar breathing, and at 1 o'clock he was cold in death. All three of these brethren we have reason to believe are to-day at God's right hand, singing praises to His name.

On the 2nd of June 1884, George W. Erambert. a former member of the Home, died at Farmville, Va., the place of his birth. His was a chequered history, full of pathos and warning. He came to the Home for the first time, two days after it was opened, and it was my privilege to give him several opportunities to combat the rum-fiend until at length he passed away forever. But his death was not that of the despairing. The last of his lapses from sobriety was brought to my notice by a note he sent to me in February, 1883, which read as follows : "For God's sake have mercy upon me. I am in rags. Do help me once more. You will find me at the Post Office." I was lying on a sick bed at the time and sent a messenger in pursuit of him, but he could not be found and I heard nothing of him until the following March, when Mr. Ottignon, a former member of the Home, called and informed me that he had brought Mr. Erambert into town with him that morning. He had walked to Mr. O.'s residence the day before (some twelve miles), and was very anxious to be received again into the Home. I had an interview with him: I could not refuse his prayer, he seemed so much in earnest, and was in such miserable health. During his last residence at the Home he had the best medical treatment at the hands of Drs. A. J. Richardson and Andrew H. Smith, but his disease (consumption) was past cure. He left us and entered the Presbyterian Hospital, receiving there the best possible care and attention. A kind friend, Mr. Case, then took poor Erambert to his home at Tremont, where he remained until March, 1884, when he went to join his friends in the South. About a year previous, in April 1883, while a member of the Home, I called him to my study and read to him a little talk which I had prepared for the Saturday evening meeting, subject: "Swept and Garnished." I shall never forget the expression of his countenance as he looked into my face and said, "That is me, swept and garnished. Oh, Mr. Bunting, will not the Holy Spirit come and take up his abode with me?" We prayed together, and as he rose from his knees he assured me that God had heard our prayer, and that he was saved. Thenceforth in his testimony he referred to this time and place as the beginning of his new life.

His was indeed a life of distress and suffering. Nights without sleep, days of weariness, gasping for breath. He struggled hard for life, determined to triumph over disease. He loved his wife and family well, and strove to bear the necessary separation patiently. In his last letter to me, written May 22d, 1884, ten days before his death, strongly clinging to his life though prostrated on bed, he said: "I am slowly improving, but am quite sick; yet I am patient, and have left all with God. His grace is sufficient for me. He alone can do for me and restore me to health. I fully trust Him. I ask your and Brother Pulis' special prayers that God may restore me to health, and that when I write you again I may be stronger." Man proposes, God disposes. On June 3rd, I received a letter from his brother. It read thus: "Dear Mr. Bunting: I write to inform you that my brother George died at half-past eleven o'clock last night. He died trusting in Christ. He spoke of you so often, and how he loved you. Poor fellow; he suffered so much. We did all we could for him. Not a day passed over his head but what he spoke of you and your kindness. How he enjoyed the letters he received from you. He said he wanted to live to go back to see you and his friends at the Home, but if God thought best to take him away, he was willing to go. I have heard him speak of you so much that I feel as if I knew you.

"Hastily yours,
"E. L. ERAMBERT."

On the same date I received a letter from his wife. She said: "George told his brother on Sunday that he could not be with them long. He spoke often of me through the day, and wished he could see me. He wanted to be spared to do something more for us, but if God thought best to take him he was willing to go. He died trusting in Christ. About one hour before breathing his last he called me by name, saying: 'Frank, I can't stay with you long.' Oh, Mr. Bunting, to think the end is come, and such an ending as this, separated as we were, and now surely separated by death. My heart is full. I can't help grieving for him. He was my husband. God forgave him; so had I. I hoped so much that I might be with him at the last. His brother said George loved you so much. Please write to me, Mr. Bunting. A letter from you will do me so much good. I feel all alone now, for while he lived there was always a hope for the future; but I shall trust as he did in Christ, and we will yet meet where there is no separation. God bless you, Mr. Bunting! you brought him at last to Christ, you taught him the way, and God has taken him home.

"Yours sincerely, MRS. F. D. ERAMBERT."

Thus died our friend and brother, at the residence of his brother in Farmville, Va. I had corresponded with Mrs. Erambert for several years previous to his death, and knew her well personally. She is a truly Christian woman, ever faithful to her poor husband, always regretting the hard necessity, the cruel habit which separated them, but ever resigned to the will of God.

I have thought it best to publish this short sketch of a life rendered profoundly wretched through strong drink, and to frankly state the many failures of a man who, no doubt, unconsciously deceived himself as to the state of his soul, until God opened his eyes to the blessed truths of salvation through our Lord Jesus Christ. The lesson of George Erambert's life is—" never despair." Christ Jesus can save to the uttermost. This poor soul, many times flung into the pit by Satan, trampled, rent and polluted, rose at last, pure and bright, triumphant through the grace and mercy of God and of His glorious Son.

Many others of our band have passed away, of whom we have confident hope that in their last hours on earth the Saviour was there to carry them through. But we only name those of whom we have positive knowledge through messages sent us.

Thurlow Weed in *The Tribune.*

The cause of temperance is being served in this city by Gospel agencies. Mr. Bunting, who has established a Home for penitent inebriates in this city, does his work thoroughly. Mr. Bunting, in his quiet, effective way, is reforming and converting what has hitherto been held to be a hopeless class of inebriates. Do not despair; there is still hope for the drunkard.

From *The Times*, New Brunswick, N. J.

Foremost among the institutions for aiding the wretched stands The New York Christian Home. It has performed its work so silently yet withal so nobly as to demand more than passing notice from the casual observer. Founded in 1877 by Mr. Chas. A. Bunting, it has so far exceeded the most sanguine expectations of its supporters. Who are the inmates? Doctors, clergymen, lawyers, actors, clerks and representatives from every craft, all brought to one common level by rum. Besides relieving those who suffer from the use of alcoholic liquors, the "Home" treats Opium and Morphine cases with equally happy results.

From *American Temperance Union.*

The Christian Home, situated at the corner of Madison Avenue and Eighty-sixth street, is not merely in name but in very deed a "Christian Home." Its pleasant, roomy and attractive apartments are not its strongest recommendations. The sympathy and love which exalt Christ, and point the victim of strong drink to Him who can save to the uttermost; that love which does not spurn the most abandoned one, but sees in him an immortal soul for whom the Son of God suffered and died—it is this love and this faith that exist there, and give to the "Home," its true name, and the blessed influence it is exerting upon its inmates and all who visit it.

From *The Congregationalist*, Boston.

The friends who have aided and watched the working of The New York Christian Home for Intemperate Men, are greatly encouraged by the outcome. The Home was opened in June, 1877. The "Home" deserves the name: it is well furnished, and has none of the air of a hospital or asylum. Forcible restraints are not used. The rules of the house are signed on entering, and their faithful observance is left to the honor of the inmates. These have for the most part been men of more than average culture and social position; members of all the professions, journalists, teachers, merchants being represented.

From *To-Day.*

* * * * There are many hundreds to-day who are thanking and praising God for this dear Home. Many of them a few years ago were wandering about our streets, poor and despised of all men, in rags and in filth, but now they are living in happy homes, surrounded by kind and loving influences, respected by all men, and they owe it all to God and The New York Christian Home. Its doors are open to all who honestly wish to lead a new life; men who are tired and sick of the wretched lives they are leading and want the mighty Saviour are welcome, the poor as well as the rich. * * No tobacco is permitted to be used by any of the members of the Home, Mr. Bunting believing this habit tends to lead men into drinking. May God bless this grand and noble work, and may He spare Mr. Bunting's life for many years to come, and may God bless every effort that is put forth to reclaim and redeem the drunkard and the lost ones.

From *The New York World.*

* * * * Mr. Bunting showed our reporter through the Institution. Everything told of a cheerful home. In the library was found an editor of a Hartford paper. He was happy and contented. In the reading-room were gathered a banker from Virginia, a former employee of the New York Comptroller's office, a former clerk for A. T. Stewart, who was at the head of the silk department at one time, an artist, a lawyer, and a resident of New York who has just run through a fortune of $150,000. In the sitting-room was a salesman in a leading house in New York, an Episcopal minister from Massachusetts, a wood-engraver who was earning $12 a day before he went to pieces. There are many wonderful cures entered on the records of the Institution. A patient came from Blackwell's Island. He was the worst they had ever received. He was bathed, clothed, and converted the same day. He was a lawyer from Ohio who had been drinking for years and had fallen to the lowest depths. His brother was a judge of the Supreme Court in Kansas. This patient is now enjoying a lucrative practice in Pennsylvania.

The following letter and reply appeared in the New York *Witness* four years ago:

Dear Editor of the Witness:

Is there any cure for drunkenness where it has been persisted in for years? I have been listening to one who, as he says, is bound hand and foot with a chain riveted at every corner. He has promised again and again, and as often broken his vows. He is now over sixty years of age, and has been drinking for fifteen or eighteen years. He knows, to his sorrow, that "It biteth as a serpent and stingeth like an adder.;" Again, I repeat, is there help? Can you suggest anything to avert the terrible calamity of seeing a loved one filling a drunkard's grave? God have mercy on those who allow this terrible curse to destroy our homes and the souls and bodies of its victims.

A ST. LOUIS YOUNG WOMAN.

[The only cure in such cases is what may be called a miracle, wrought by God in answer to the prayer of faith. There have been many instances of such cures of habitual drunkards. The least yielding afterward to taste any kind of intoxicating drinks will renew the uncontrollable drunkard's appetite. The Christian Home for Inebriates, Madison Avenue and 86th Street, New York, has saved many apparently hopeless drunkards by teaching them to lean on the arm of the Saviour continually.—ED.]

From the *Telegram*, N. Y.

" Don't tell me there is no hope for a drunkard,' said a gentleman to a *Telegram* reporter in front of the post-office. " Do you see that finely dressed man there?" pointing to a handsome looking specimen of humanity who was conversing with a lady on the corner. " Less than one year ago there could be found fewer more miserable creatures in New York than that man. He was dressed— almost undressed—in rags, and he ' hung out' in some of the lowest dives on Chatham, Roosevelt, and Mulberry Streets. When he couldn't get six-cent rum he was only too glad to get a three-cent glass of Italian made 'lightning,' or even a two-cent schooner of stale beer. His brother, a very estimable business man, had cast him off. He was friendless and in despair. He became a loathsome beggar. By some miracle he obtained admission to The Christian Home for Intemperate Men up town and remained there five weeks. He was fed and clothed for some time, and when he left was a reformed man. He now holds a responsible and lucrative position down town ; he has been restored to his family and friends, and to my certain knowledge has not touched liquor, gin or beer for eleven months, and I do not believe he could be induced to do so for $10,000.

" Yes, some drunkards can be reformed, for I have been a drunkard myself— but not in the past fifteen years."

From *The New York Times*.

The good results which have been effected at The New York Christian Home for Intemperate Men is shown by the testimony of former and present members. One, a Presbyterian minister's son, was reformed by reading a tract at the Home. Another, a ruddy-faced, bright-eyed man, never read a line in the Bible until he came to the Home. He sank so low that he lost all self-respect, and even slept in lumber piles on the docks. While intoxicated one day, a gentleman who saw him wrote the location of the Home on the margin of a newspaper without giving its name or object, and told him to go to the address given. He forgot about the scrap of paper until the next day, when he unthinkingly pulled it out of his vest pocket. A third, who two weeks ago was as drunk as whiskey could make him, said he had become a Christian man and lost his taste for liquor. A fourth, a fine looking man, was induced to go to the Home by a person who was reformed by its inmates. An outcast for ten years who had been in prison, said he came to his senses at a noonday prayer-meeting, and was directed to the Home, which completed his reformation.

From *The Christian Advocate.*

"Where shall we send an intemperate man who wishes to reform?" How often this question comes to us by mail and in conversation! There may be many good places: we hope there are. But before recommending any, several things must be known. Does the guilty man wish to reform? Does he feel guilty or only unfortunate? Does he excuse himself in his debaucheries because he is diseased? Does he feel heedless without Divine aid? If these questions can be answered in the affirmative, we have no hesitancy in answering by a positive recommendation the question so often asked. Here in the city of New York is the place where we would put an intemperate son, brother or friend with the greatest hope. It is The New York Christian Home for Intemperate Men, corner of Madison Avenue and 86th Street. Its results are wonderful. Moral miracles have been wrought there, and some of our own acquaintances have been thoroughly reformed. With emphasis we repeat our former statement, that if we had an intemperate relative or friend anywhere who wished to reform, we should spare no pains or expense to get him under the power of this Christian Home. Charles A. Bunting is the Resident Manager.

"Ebenezer" in *The Tribune.*

* * * * It is of recent date that the writer became acquainted with the workings, the object and methods of this Institution, and yet for some ten years past it has been performing a beneficent mission in this community, restoring manhood to the unmanned, peace to families where discord had long reigned, hope to despairing souls and joy to hearts long saddened with the burden of reckless sin. The care of the intemperate, their cure and restoration to respectable citizenship, is an object well deserving the support of the philanthropist. But when is added the regeneration of the man and his elevation to a place in the ranks of Christian civilization, then the work appeals to all who claim spiritual affinity with Christ. It is precisely this special kind of work which is the self-chosen duty of those who manage and conduct this Institution, and when it is considered that the great majority of those who enter to share its blessings leave its precincts saved men, men redeemed from the curse of alcoholism and restored to conditions of usefulness for society, words fail to fittingly describe its peculiar merits and its claims upon the community.

The names of those who are found among the most strenuous supporters of this Home, suffice to prove how completely it has fulfilled its mission, and it only remains to be generally known that the late William E. Dodge was among the first founders of this work to assure its advocates an attentive hearing from earnest Christian men and women. His son, the Rev. D. Stuart Dodge, occupies his deceased father's place as President of the Institution, and under his watchful supervision it pursues its great object, leading men to know God, and because of that knowledge to abandon their degrading vices. To support such a work is, I conceive, one of the highest forms of Christian benevolence, and a visit to the Resident Manager, Mr. Charles A. Bunting, will amply satisfy any one desiring to forward one of the most effective forms of Christian work, that in truth and in deed is God wonderful in all things, even to the salvation of those whom society so often dooms to utter degradation and destruction.

[The New York *Herald*, *The Sun*, and other leading papers, have published similar commendatory notices of our work.]

ABSTAIN FOR THY BROTHER'S SAKE.

The following is a portion of one of the weekly "talks" or lessons, given by the Manager on either Tuesday or Saturday evening, the subject being taken from the 14th and 15th chapters of Romans:

According to the teachings found in God's Word, the strong should be willing to bear the infirmities of the weak, and not to please themselves. If God in His kind providence has been so good as to take our feet from the miry clay and put them on the solid rock, we should aim to do all we can for those who are as yet under bondage. We should not be selfish. If we are able to drink our ale and wine, and find we can do all this in moderation, and yet *know* that our example is such that others are led to ruin by it, should we not do as He did—"for even Christ pleased not Himself." Let us deprive ourselves of these so-called luxuries for the sake of our weak brother. Self-denial, as to personal gratification, will be found pleasing to God, and our joy will be increased as we find we are enabled to live in this way for His sake. As our minds are enlightened more and more every day, and we see the pernicious example of some of those who are recognized as Christian people, should we not be bold as Gospel temperance workers in denouncing everything which has a tendency to endanger our brother? "It is good neither to eat flesh, nor to drink wine, or anything whereby thy brother stumbleth, or is offended, or is made weak." Duty requires us to abstain from indulgences which may lead others to sin. It is for us who have obtained the knowledge to preach it by example as well as by word. Let us "who are of the day be sober, putting on the breastplate of faith and love." Let us not live in stupidity, unmindful of the great truths which are every day presenting themselves to our view. Let no uncertain sound be given to this Gospel trumpet. Let us not be led to believe that the man who has the capacity to drink, and not show it either by conversation or other outward appearance, is any less free from sin in God's sight than the one who cannot drink without plainly showing its accursed effects. "God is not mocked." His eye can see these things, though they may be hidden from the eyes of the world. He will not look upon sin in any form with the least allowance. We should by our example try to strengthen our weaker brother by denying ourselves in every way that we can. Our object in trying to live blameless before men should not be for the sake of merely our own selves, but for the good such a life exerts. Oh, how pleasing in God's sight it will be for the man or woman who for Christ's sake abstains from these indulgences. No one should dare to take the risk in doing anything upon which he cannot ask God's blessing. He who doubts the lawfulness of anything, and yet does it, when there is no doubt about the lawfulness of abstaining from it, is condemned as guilty of sin. "We then who are strong ought to bear the infirmities of the weak, and not to please ourselves."

ABSTAIN FOR THY BROTHER'S SAKE.

If we are not satisfied as to the right of our using tobacco, and especially after having heard the personal danger of such use from the lips of those who have returned to drink from the use of tobacco—that it has proved the cause of their falling—should we dare to trifle with it? God forbid! Again, have we a right to take that which we have consecrated to the service of our Master and burn it up? Are we honest Christians when we take that which does not belong to us and use it for our own selfish desires, when perhaps at that very moment some poor woman or orphan child may be suffering for bread? Or again, have any of you a right to use your money for tobacco after going from this Home owing for your board while you were here? How can we satisfy our consciences while doing these things? Many a promise like this has been made in the Home (when it has been our privilege take the poor sufferer into the Home from the street), "I will, if ever able, pay for my board." Has any such person the right to indulge in his so-called innocent luxuries until he has liquidated such a debt?

Oh, let each one of us strive to live more like the apostle Paul. "Wherefore," he says, "if meat make thy brother to offend, I will eat no flesh while the world standeth, lest I make my brother to offend." "Let each one please his neighbor for his good to edification." For even Christ pleased not Himself, but as it is written, "the reproaches of them that reproached thee fell on me."

SWEPT AND GARNISHED

(Luke xi.: 21-27.)

The following is the lesson referred to in the sketch of George W. Erambert, published in the "Obituary Notices," and which led that poor soul to at last know and feel the wonderful grace and boundless efficacy of Christ's salvation.

Many hours have I passed since I entered upon the Christian work of this Home, pondering over one of the most pertinent of questions. Many times have I asked the question of those much more advanced in the Christian life than myself, but invariably have I received an evasive answer. "Tell me, if you can," has been my question, "how it is so many fall back into their old sinful lives, who gave such apparently unmistakable evidence of a new life commenced—of being "born again?" Sometimes for weeks, yes. for months, the great change will be apparent to all, and even the ungodly begin to say, "Yes, of a truth there is power to transform in this religion ." when perhaps. even the unbeliever's heart has begun to ask the question of himself, "Why do I delay longer? Why procrastinate? Why not now secure to myself this saving power?" Lo, and behold! in a moment, in the words I have read you. he, the one on whom all eyes have been fixed, and perhaps the destinies of so many have been placed, "taketh to him seven other spirits more wicked than himself, and they enter in and dwell there : and the last state of that man is worse than the first."

Why this state of affairs? How can these things be? Is there no balm in Gilead ? Is there no physician there? Will not God's grace *keep* as well as save for a time? Oh! blessed be His holy name! yes; ten thousand times yes! shall be the reply. God not only saves, but he keeps to the very uttermo t *all*, yes, *all*. I want to emphasize that little word, *all*. In Hebrews viii. ch., 10 v., you will find these words: "For this is the covenant that I will make with the house of Israel, after those days. saith the Lord : I will put My laws into their mind, and write them in their hearts, and I will be to them a God, and they shall be to me a people, and they shall not teach every man his neighbor. and every man his brother, saying know the Lord : for all shall know Me from the least to the greatest. For I will be merciful to their unrighteousness and their sins : and their iniquities will I remember no more,"

A *covenant* God has entered into, and His is a sure covenant. And mind you, it is entered into only with the house of Israel ; and if we are members of that household, God's word for it, it can never be broken.

Now, once more I ask the question, How is it that so many fall back again into their old sinful lives, after giving such unmistakable evidences of a change of heart? It is now no longer a mystery ; it is solved, and I think for the first time I know what my Master intended I should know, clearly. In His covenant He says, "I will write my law in their hearts." This being done, God's grace promised is made complete, for the judgments and affections together command the man. In reading a book a short time since, upon this subject, the whole thing appeared in clearness as does the noonday sun.

The judgment *alone* may at times be convinced in an unregenerated person, if not of the *infinite excellence* of God's law, yet he may be convinced of the folly and danger of opposition to it; and the effect of this, under favorable circumstances, such for instance as being in this Home, sitting daily under its teachings, a very considerable reformation may take place, insomuch that he may be led to say, "I feel a great change." The desire for bad habits, bad company, and the outward conduct in many particulars may be greatly changed, and will lead the man to say, "My desires for intoxicating liquors and tobacco are all gone." Satan may leave a person under such circumstances, and go out for a season, and that man's house may have been swept and garnished, as I have read to you from God's Word, and in all sincerity the man may have been led to say, "I know I am safe," when the fact is that, although the house may have been swept and garnished, it has been all the time empty. And as you will see, God's grace never having occupied it, Satan is at liberty to return and occupy it whenever he pleases. So when its surroundings change we find the affections are yet upon sin, and thus sooner or later the soul is drawn into sin against its own best judgment and clearest dictates of conscience. Here we see the deception of our hearts. Honest in every purpose, yet failing to grasp the great truths of God's law. For let me ask you, my brother, What is God's law? Thou shalt love. Do we? Is there a pleasure and a liberty in the service of God? Do we take pleasure in his commands? Are they at all grievous to us? If they are, let us suspect our religion. Let us suspect our interest in the grace of this covenant.

Religion pure is the delight of God's service. Religion is the love of God. The obedience of the true believer is an obedience not of constraint, but of free and hearty choice.

First.—No drug or medicine of any description shall be brought into the house, or used by any member, without the knowledge and consent of the Manager.

Second.—All members are expected to remain in the Home until, in the judgment of the Manager, they are fitted to go forth.

Third.—Each member is expected to aid cheerfully in this work of reformation by courteous deportment, cleanly habits, and a willing acquiescence in rendering such service as he may be called upon to give, thus tending to make this a happy home.

Fourth.—Unbecoming language and heated discussions on religious or political subjects, that may lead to strife or dissentions, cannot be allowed, and all allusions to the past life of sin are strictly forbidden.

Fifth.—Meals will be served at regular hours. Regularity and punctuality are absolutely required of all members.

Sixth.—Members of the Home are required to be present at the regular evening services; also at the morning and afternoon meetings and evening prayers.

Seventh.—The use of tobacco is strictly prohibited.

Eighth.—Every member of the Home is expected to take a bath once a week, for which extra towels are furnished.

Ninth.—Members guilty of drinking, using tobacco, or bringing liquor into the Home, shall be summarily dealt with, as this is so gross a violation of obligation and honor that it cannot be overlooked. Others knowing of such conduct, and not reporting the same to the Manager, become parties to the wrong, and cannot expect to enjoy the confidence of the management.

Tenth.—While a system of petty watching and of making complaints about trifling matters is deprecated, justice to the interests of the Home demands that any grave offence or violation of rules shall be brought to the attention of the Manager.

Eleventh.—All members are expected to be ready to retire at the hour designated, in order that the doors may be closed, lights extinguished, and the house quiet by said time, that those desiring repose may not be disturbed. And after retiring to their rooms, conversation is strictly prohibited.

ACT OF INCORPORATION.

ACT OF RE-INCORPORATION.

CONSTITUTION AND BY-LAWS.

During the past year we petitioned the Legislature to make certain material modifications in the above subjects of legislation. It was our hope to have had these changes ready for publication before this book went to press ; but as the matter has been delayed at the Capitol, we find it necessary to request that any one desiring information in connection with the actual legal status of The New York Christian Home, would please apply for the same to the Resident Manager.

CONTENTS.

FRONTISPIECE [Mr. Bunting.]
DEDICATION

	PAGE
PREFACE	5
ORIGIN OF THE WORK	7-14
My own Conversion	7
Rev. Dr. Tyng Lends Assistance	8
God sends Wm. E. Dodge	9
Our First Meeting	10
First Public Appeal	11
Securing a Location	12
The Home Opened	13
Photo-Engraving of old Home	13
OBJECT AND PLAN OF WORK	14-22
The Gospel Remedy	14
God's hand with us	15
Helping One Another	16
Drunkenness not Hereditary	17
Conditions of Admission	18
Looking to Jesus only	19
Old Friends Remembered [Wm. T. Booth and others]	20
Words of Arthur W. Parsons	21
A Source of Public Economy	22
THE NEW HOME	23-32
Photo-Engraving of New Home	23
New Quarters Opened	23
The Building Fund	24
Mr. Dodge's Sainted Memory	25
Generous to the last	26
His Loss Mourned	28
Photo-Engraving of the late Wm. E. Dodge	28
Steadfast and Tireless Workers. [Rev. D. Stuart Dodge, Caleb B. Knevals and others	29, 30
List of Officers and Directors	31, 32
FRUITS OF THE WORK	33-42
Converting a Scoffer	33
"The Gospel Temperance Salesman"	34
A Marvellous Conversion	35
Rev. Dr. J. M. King speaks	36
Outsiders Brought to Jesus	37
Accepting Salvation	38
A Morphine Slave Redeemed	39
Physically and Spiritually Transformed	40
Dr. H. C. Houghton's Views	41
FACTS FOR CHRISTIAN THINKERS	42-87
The Drunkard Needs Sympathy	42
A National Influence	43
The Bible Our Text-book	44
Tobacco a Snare	45
Tobacco Habit Involves Drinking Habit	46

	PAGE
FACTS FOR CHRISTIAN THINKERS	42-87
Telling Testimonies	47
Steadfast for Years	48
Confessing Christ	49
Workers Wanted	50
Appeal to Christian Women	51
A Tribute to Mrs. Bunting	52
Evils of Moderate Drinking	53
Tippling in Christian Homes	54
The High License Fallacy.—No Compromise with Satan	55, 56
License Crime, and Ruin Your Boy	57
Suppress it Utterly	58
Rum the Common Enemy.—The Rumseller Plunders the Home	59, 60
Rum-selling a Crime	61
It is Our National Sin	62
Woe unto Mighty Drinkers	63
The Drunkard's Terrible Appetite	64
Bartering All for Rum.—Soul and Body for Gin	65, 66
The Morphine or Opium Habit	67
"Literally Created by a physician"	68
Mendacious and Dishonorable	69
How to Cure a Morphine Fiend	70
The Cocaine Habit.—Its Awful, Deadening Slavery	71, 72
The Chloral Habit	73
A Stay Here Nourishes the Soul	74
"A Hard Saying; Who Can Hear It?"	75
The Spirit Quickeneth	76
To Friends of the Intemperate	77
A Monument of Mercy	78
Photo Engraving of J. L. Pulis	79
My Co-worker Speaks	79
Describing Our Spiritual Work	80
Can the Drunkard be Saved?	81
Mr. Pulis on the Tobacco Habit	82
God's Keeping Power	84
How the Drunkard is Saved	85
A Consecrated Will	86
Visit Us and See	87
SUPPLEMENTARY MATTER	88-124
Statistical Tables	88
Letters from Former Members	89
Anti-Tobacco Items	100
Bequests	103
Vanderbilt Fund [a munificent legacy]	103
Acknowledging Cornelius Vanderbilt's Kindness	104
Memorial Fund	104
Obituary Notices	105
Newspaper Notices	111
Abstain for thy brother's Sake [a "Home" lesson]	116
Swept and Garnished [another "Home" lesson]	118
Rules, By-laws and Constitution, Act of Incorporation, &c., &c.	120, 121

www.ingramcontent.com/pod-product-compliance
Lightning Source LLC
Chambersburg PA
CBHW020108170426
43199CB00009B/446